Volume 3 of the American Wild

THE PACIFIC CO
WILDLIFE RE

REVISED EDITION

By Charles Yocom, Ph. D.
Professor of Game Management

and

Raymond Dasmann, Ph. D.
Chairman, Division of Natural Resources
of Humboldt State College, Arcata, California

Edited by Vinson Brown

Illustrators
Plants: Charles Yocom
Animals: Elizabeth Dasmann

Copyright © 1957, 1965, by Charles Yocom
and Raymond Dasmann

Paper Edition   ISBN 0-911010-04-1
Cloth Edition   ISBN 0-911010-05-X

TABLE OF CONTENTS

Introduction 3
Key to Plant Parts 6
Common Plants 7
    Coastal Strand 7
    Coastal Marshes 14
    Coastal Brushfields 17
    Coniferous Forests 24
    Coastal Hardwood Forest 45
    Woodland-Prairie 50
Common Animals 55
    Mammals 55
    Birds 76
    Reptiles and Amphibians 103
    Fishes 115
Suggested References 117
Index 117

Published by Naturegraph Co., Healdsburg, California

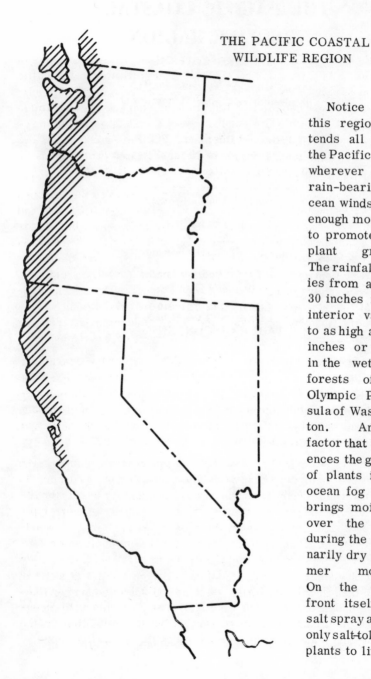

THE PACIFIC COASTAL
WILDLIFE REGION

Notice that this region extends all along the Pacific Coast wherever the rain-bearing ocean winds bring enough moisture to promote lush plant growth. The rainfall varies from around 30 inches in the interior valleys to as high as 120 inches or more in the wet rain forests of the Olympic Peninsula of Washington. Another factor that influences the growth of plants is the ocean fog which brings moisture over the land during the ordinarily dry summer months. On the ocean front itself the salt spray allows only salt-tolerant plants to live.

INTRODUCTION

This book is the third of a series on the wildlife regions of America. Wildlife regions, such as the Pacific Coastal Wildlife Region, are distinctive natural geographic areas of similar climate and topography, which tend to have characteristic animals and vegetation within their boundaries. Overlapping between regions makes it impossible to draw a rigid line separating them. Some species of plants and animals appear in several regions.

The Pacific Coastal Wildlife Region, as defined in this book, extends from Monterey County, California, north to the southern part of British Columbia. Farther north, in coastal Canada and and Alaska, similar plant and animal life may occur, but, in addition, various northern species are found which are not included in this book. The famous California redwoods are the dominant trees of the southern part of the region. Farther north Sitka spruce, cedar and hemlock join with Douglas fir to form magnificent forests renowned for their fine lumber. Living in this region are a variety of interesting and unique animals, ranging from the majestic Roosevelt Elk to the tiny Red Tree Mouse.

The purpose of this book is to acquaint you with the characteristic plants and animals of the region. Since the book has been kept small to make its selling price reasonable, only basic information is given about the plants and animals most likely to be encountered. For more details, the reader should seek the larger and more complete books listed in the bibliography.

In the first section of this book the plants are grouped in the vegetation communities in which they most commonly occur. In the coniferous forest, which occupies the greatest area in the region, certain species are typical. Other species are more typical of other habitats, such as the coastal strand, or the woodland prairie. Photographs of these plant communities or habitats are shown at the beginning of each habitat section. Following the descriptions of the plants of each habitat is a list of the common mammals, birds, reptiles and amphibians most likely to be encountered in that vegetation type. Learning to recognize the plant communities, vegetation types, or habitats is the first step in learning to recognize the animals that live there.

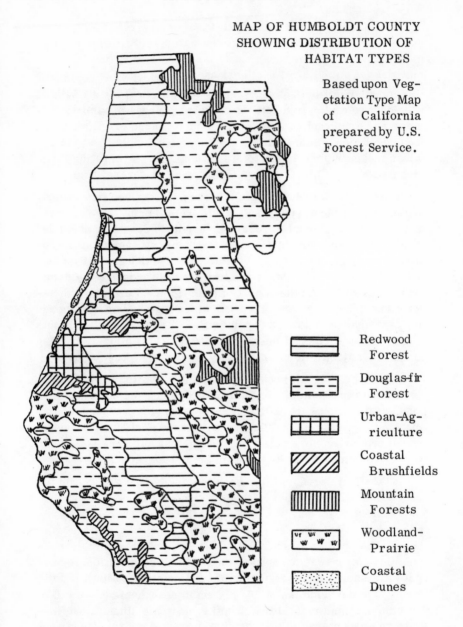

MAP OF HUMBOLDT COUNTY
SHOWING DISTRIBUTION OF
HABITAT TYPES

Based upon Veg-
etation Type Map
of   California
prepared by U.S.
Forest Service.

Redwood
Forest

Douglas-fir
Forest

Urban-Ag-
riculture

Coastal
Brushfields

Mountain
Forests

Woodland-
Prairie

Coastal
Dunes

The "mountain forests" shown on this map are dominated by
ponderosa (or yellow) pine and actually belong to a different wildlife region, the Sierra Nevadan Wildlife Region (see volume 2).

In later sections of the book the mammals, birds, reptiles, amphibians and fish most likely to be encountered within the region are illustrated and described. The reader who finds an animal species listed for a particular vegetation type, should then turn to the appropriate section of the book to find the illustration and description for that species.

PACIFIC COASTAL REGION WILDLIFE HABITATS

If you were to travel inland from the seacoast in the vicinity of Eureka, California, to the higher mountain ranges of the interior, you would pass through a series of vegetation types, such as are shown in the map on page 4 and illustrated in the habitat sections: coastal strand, coastal marshes, coastal brushfields (or northern coastal scrub), coniferous forests (including both redwood forests and spruce and fir forests), hardwoods, and finally, farther inland, the woodland-prairie area, with its groves of oaks interspersed with grasslands. All of these habitats may not be encountered in any one section through the coastal region, or their arrangement may vary from one location to another.

Another type of habitat, important in most wildlife regions, is the streamside woodland. But in this relatively well-watered coastal region, the streamsides are usually not characterized by distinctive vegetation, and are therefore included, for the most part, with other vegetation types. Man-created habitats, found around cities, towns and in agricultural areas, also are not described separately, but are, nevertheless, important areas where wildlife is found. The diversification of vegetation that occurs near human habitations often favors the presence of larger numbers of animals, particularly birds, than will be found in the adjoining, less disturbed areas. So forest birds will move into the parks and wooded gardens, while prairie or grassland birds and mammals will visit vacant lots, fields and the more open gardens.

To aid in understanding identification and distribution, names of habitats are marked on the margin of each description as follows: Strand = coastal strand, Marsh = coastal marsh, Brush = coastal brushfields, Conif. = coniferous forests, Hardw. = hardwood forests, and Wd. Pr. = woodland-prairie. Besides these main habitats (all pictured in the book) some other habitats mentioned include: Meadow = forest meadow, Water = fresh and salt water, Str. Wd. = streamside woods, Rocks = rocks, Oak = oaks.

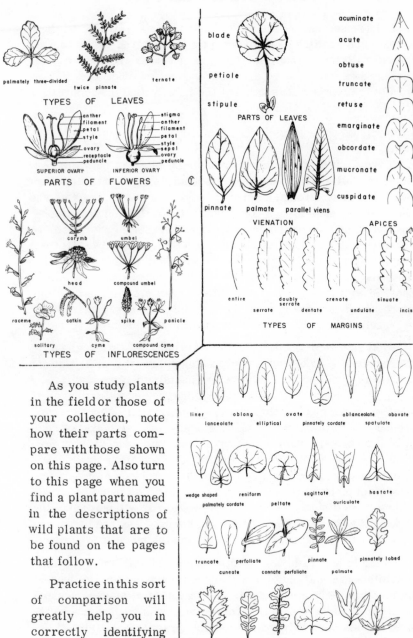

palmately three-divided

twice pinnate

ternate

TYPES OF LEAVES

anther
filament
petal
style

stigma
anther
filament
petal
style
sepal
ovary
peduncle

ovary
receptacle
peduncle

SUPERIOR OVARY

INFERIOR OVARY

PARTS OF FLOWERS

blade

petiole

stipule

PARTS OF LEAVES

pinnate   palmate   parallel viens

VIENATION   APICES

acuminate

acute

obtuse

truncate

retuse

emarginate

obcordate

mucronate

cuspidate

corymb

umbel

head

compound umbel

raceme   catkin   spike   panicle

solitary   cyme   compound cyme

TYPES OF INFLORESCENCES

entire   doubly serrate   crenate   sinuate

serrate   dentate   undulate   incised

TYPES OF MARGINS

As you study plants in the field or those of your collection, note how their parts compare with those shown on this page. Also turn to this page when you find a plant part named in the descriptions of wild plants that are to be found on the pages that follow.

Practice in this sort of comparison will greatly help you in correctly identifying the common plants you find.

liner   oblong   ovate   oblanceolate   obovate
lanceolate   elliptical   pinnately cordate   spatulate

wedge shaped   reniform   sagittate   hastate
palmately cordate   peltate   auriculate

truncate   perfoliate   pinnate   pinnately lobed
cuneate   connate perfoliate   palmate

pinnately cleft   pinnately divided   palmately three-cleft

pinnately parted   palmately three-lobed   palmately three-parted

TYPES OF LEAVES

COMMON PLANTS AND
WILDLIFE HABITATS

In the following pages only the more common and most frequently encountered species are described, the few exceptions
being those plants that are of special interest. A complete listing
of the plant species alone would take a book many times this size.
In order to identify the plants, illustrations and brief descriptions
are given. It will aid you in understanding the descriptions if you
study the illustrations on page 6 and learn to recognize the shapes
and kinds of plant parts.

Ocean shore, Humboldt Co. Photo reproduced by permission
of the Redwood Empire Association.

PLANTS OF THE COASTAL STRAND AND BLUFFS

Sand dune areas and nearby bluffs are everywhere characterized by distinctive vegetation. Because sand has a low capacity
for holding water, many of the plants found along the coast will
be closely related to species found in arid habitats, such as dry
high mountain or desert regions, farther inland. Coastal strand
vegetation varies from low growing succulents along the high tide
line, through grasslands, brush areas (dominated by bush lupine

and other shrubs), to well-developed forests in which the beach and Monterey pines are characteristic.   The great variety of vegetation is very favorable to birds of many kinds.

BEACH PINE (Pinus con-
torta).   This scrub pine, which seldom grows over 35' high, has comparatively thick branches, forming a *Strand* round-topped, compact and symmetrical, or an open, picturesque head. The ovoid to sub-cylindric cones are usually 1-2'' long and remain on the tree from 4 to 5 years or more.   This attractive seacoast tree, which has lit-tle commercial value, ranges from Mendocino Co.   north along the coast to Alaska.

Strand

← MONTEREY   PINE (Pinus radiata). A 25 to 75' or higher tree, usu-ally well-shaped, but sometimes with a flat-tened or broken top when old.   The bark is dark, rough, hard and often deeply fissured; the dark green needles appear in 3's, rarely in 2's, and are 3-6'' long; the cones are light or reddish-brown in color and 2 1/2 to 4'' long. It is found in a few scattered colonies from Santa Cruz Co. S.

Strand BISHOP PINE (P. muricata) is found from Humboldt Co. south, and has very long (4-6'') needles in 2's; cones often with spurs.

BISHOP PINE (Pinus
muricata). A small tree
usually not over 60' tall.
The bark is dark, purplish
brown, deeply furrowed
and scaly. The needles
are 3'' to 5'' long, yellow
green, in dense clusters
at the branch ends. Both
the male and female flow-
ers are clustered at the
ends of the branches. The
mature cones are 2'' to
3 1/2'' long, unsymmetric
with large black seeds.
The tree is too small for
commercial use, though it
is widely used for fuel.

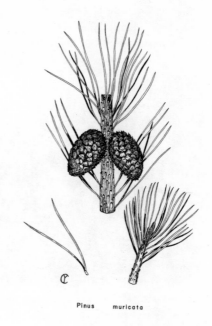

Pinus   muricata

MONTEREY CYPRESS
(Cupressus macrocarpa). A
15 to 80' tree, with tiny,
scale-like and overlapping,
grayish-green leaves; the
small cones with brown
seeds; bark dark brown.
These unusual trees grow in
dense stands only in restricted localities on the headlands
around the mouth of the Car-
mel River in Monterey Co.,
California. A very similar-
looking tree (though sometimes a low shrub), is the
MENDOCINO CYPRESS (Cu-
pressus pygmaea) of the
Mendocino coast, which has
black instead of brown seeds.

Strand

There is also the GOWEN CYPRESS of the Monterey coast, which
grows only from 1 to 5' high and has a smooth, compact crown.

Stand of Monterey Cypress on cliffs at Cypress Point, California. Photo courtesy of Southern Pacific Company.

WAX MYRTLE (see p. 18).

Strand

TREE LUPINE (Lupinus arboreus). From Humboldt Co. S. to Santa Barbara Co. this bushy lupine makes the beaches beautiful during the late spring and early summer when the bright yellow flowers are in bloom. The seeds, born in a typical bean-like pod, furnish food for quail and other seed-eating birds and mammals. The bushy plants are often 3-4' tall and spread out over several square feet of area. Several other species of lupines, usually with blue flowers, are found along the coast. Two lupines common on

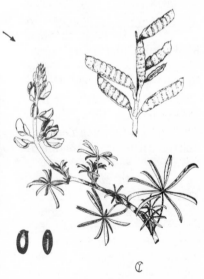

the Oregon and Washington coasts are the BICOLORED LUPINE (L. bicolor), with broad petals, one of them purple and white (this species is also in California); and the SEASHORE LUPINE (L. litoralis), which lies flat on the sands.

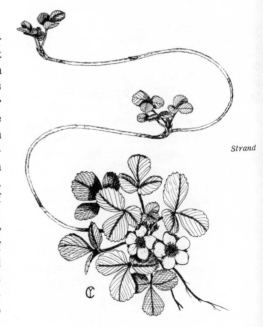

← BEARBERRY (Arcto-staphylos uva-ursi). A trailing, prostrate plant, with shiny, glabrous, oval or obovate, 1" long leaves. Common among the dunes from Mendocino Co. N. to Alaska. In some areas it is common under the beach pines. It has white or pinkish flowers, and brilliant red or pink berries, often eaten by animals and Indians. The smooth, dark red bark is distinctive.

Rocks Strand

SAND STRAWBERRY (Frageria chilensis). This plant is an important sand-binder in the coastal dunes and extends numerous prostrate stems or runners. The relatively large white flowers are carried in cymes. The small strawberries are edible and are eaten by many birds and mammals. The upper surface of the leaf is dark green & shiny smooth; the under surface is densely hairy, as are other parts of the plant. This plant is found along the sand dunes, beaches and cliffs from San Luis Obispo Co., California, north to Alaska.

Strand

SAND VERBENA (<u>Abronia</u> <u>latifolia</u>). This is a common seaside plant in the sandy soil and among the dunes, which it helps to hold in place with its long roots. It has showy, yellow flowers that are borne at the end of stalks in a many-flowered head. The stout stems of this prostrate, creeping plant are from 1' to 2' long. The herbage is very juicy and glandular; the leaves are opposite. This species is found along the coast from Santa Barbara, California, N. to British Colum.

Strand

LIZARD-TAIL (on page 22).

SEASIDE PAINTED CUP (see page 22).

COAST GOLDENROD (<u>Solidago</u> <u>spathulata</u>). In the fall, from Coos Co., Oregon, south, this plant's yellow flowers, borne in a spike-like panicle, brightens the sand dunes and hills all along the sea. The usually smooth-surfaced leaves are spatulate to obovate and toothed beyond the middle. Several 15-18" high stems rise from a creeping base. The flower heads usually have 5-10 quite small ray flowers. Indians ate the young leaves in salad.

Strand

SEASIDE DAISY (page 22).

COMMON ANIMALS OF THE COASTAL STRAND

Mammals
Bobcat, 59
Raccoon, 61
Long-tailed weasel, 63
Short-tailed weasel, 63
Striped skunk, 64
Black-tailed jackrabbit, 65
Brush rabbit, 65
Pocket gopher, 67
White-footed mouse, 71
Meadow vole, 71
Harvest mouse, 71
Bats, 75

Reptiles & Amphibians
Western fence lizard, 104
Western skink, 104
Foothill alligator lizard, 104
Gopher snake, 107
Common garter snake, 107
Western toad, 114

Birds
Turkey vulture, 81
Duck hawk, 82
Sparrow hawk, 82
California quail, 84
Killdeer, 84
Long-billed curlew, 85
Willet, 85
Dowitcher, 86
Gulls, 86
Swallows, 92
Raven, 93
Wren-tit, 94
Pygmy nuthatch, 94
Bewick wren, 94
Myrtle warbler, 98
White-crowned sparrow, 101
Golden-crowned sparrow, 101
Song sparrow, 102

14

COASTAL MARSHES

Two types of marshes are encountered along the coast. (1) The salt-water marsh is characterized by extensive tidal flats. Below the average low tide mark are found fields of EELGRASS (Zostera marina). These are easily recognized by being largely submerged maritime herbs with very narrow, grass-like leaves, 1-4" long and sheathed at their bases. The flowers have no petals, but appear as simple, single stamens or pistils on flattened, fleshy spikes. Eelgrass is an important food for the black brant.

Marsh

Higher parts of the salt water marsh are characterized by PICKLE-WEED (Salicornia), a low, very juicy-stemmed herb with jointed stems and scale-like leaves; and SALT GRASS (Distichlis spicata),with rigid, erect stems, surrounded usually by two rows or ranks of stiff leaves, and topped by short, dense, few-flowered panicles. These marshes usually then give way to grasslands or to

other upland vegetation.

Water Marsh

Photo shows fresh water marsh with partly- submerged AR-ROWHEAD PLANTS (Saggitaria), with 4-15" arrow-head shaped leaf and small, white, 3-petaled flowers. Indians

Photo by J. A. Munro, courtesy Canadian Wildlife Service.

boiled or baked the roots like potatoes, hence "Indian potato" is a name sometimes given to this plant.

(2) The _fresh_ _water_ _marsh_ is characterized by submerged aquatic plants, such as pondweeds and arrowheads, a few floating plants, such as the water lily, and, toward the shore line, emergent plants, such as the cattail and bulrush. The marshes form an important habitat for many birds and mammals.

COMMON CAT-TAIL (_Typha_ _latifolia_). Not illustrated. This common marsh plant, which often grows from 3-6' tall in the coastal area from California to Alaska, is well-known to most people. The attractive cat-tails, or thick brown flower spikes, are often used in decorating homes in the fall. The leaves were used in basket-making by the Indians. These leaves arise long, stiff and grass-like from the shallow water. The plant is an important food for muskrats and other aquatic mammals and birds.

Marsh

SKUNK CABBAGE (_Lysichitum_ _americanum_). This is a showy, perennial, glabrous (hairless) herb with large leaves, and each spike of flowers of a single sex (unisexual). The flowers are crowded onto a spadix (thick spike) surrounded by a yellow spathe (clasping leaf). The basal leaves are from 1-4' long and usually about 1' wide, rising from a stout rootstock. Skunk cabbages are common in fresh water marshes from the Santa Cruz Mts. of California north to Alaska. Indians sometimes used this plant for food.

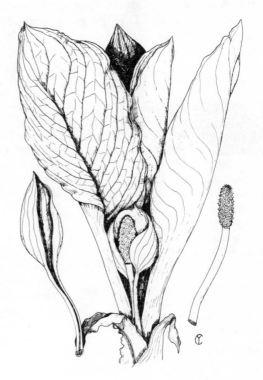

Conif. Marsh

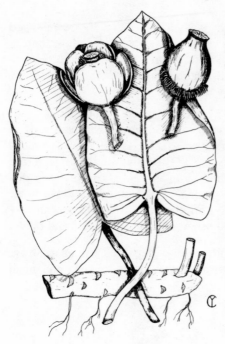

YELLOW WATER LILY (Nuphar polysepalum). The large, heart-shaped leaves of this plant, 6 to 12" broad and 6 to 14" long, are very distinctive. They are attached to a creeping rootstock by means of a long stem or scape. This is a common plant in many of the lowland marshes and lakes along the coast from Santa Cruz Co., California, to British Columbia. It is also found in lakes at higher elevations. Indians used this plant for food, and it is also eaten by muskrats and waterfowl.

COMMON ANIMALS OF THE COASTAL MARSHES

Mammals
Raccoon, 61
Mink, 62
Otter, 63
Beaver, 68
Muskrat, 69
Meadow vole, 71
W. jumping mouse, 71
Shrew-mole, 73
Vagrant shrew, 73
Most bats, 74

Reptiles & Amphibians
Pacific pond turtle, 108
Common garter snake, 109
Western garter snake, 107
Tiger salamander, 110
Frogs (except yellow-legged), 114
Western toad, 114

Birds
Grebes, 77
Great blue heron, 78
American egret, 78
Black-crowned night heron, 78
Marsh hawk, 82
Duck hawk, 82
Canada goose, 79
Ducks, 79-80
Virginia & sora rails, 84
American coot, 84
Shore birds (except willet), 84-86
Yellowthroat, 97
Red-winged blackbird, 98
Brewer blackbird, 98
Song sparrow, 102

Photo of coastal brushfields between Eureka & Crescent City, California, taken by Gabriel Moulin Studios of San Francisco.

On mountain slopes facing the ocean, where climate does not permit forests to develop, dense brushfields are frequently encountered. Similar habitats develop farther inland where soil or rock conditions do not favor forest development. Logged or burned areas of forest also spring up temporarily into stands of dense brush which are only slowly replaced by forest trees. In these burnt over areas, however, and especially in the mountains, trees, shrubs and plants of the hardwood forests (see page 45) may temporarily take the place of the coniferous forests.

These brush fields often superficially resemble the chaparral of interior regions, but typically consist of different species and are more luxuriant and often more difficult for men to break through. They give safe hiding places for many animal species.

COW PARSNIP (Heracleum lanatum). Illustrated p. 18. 4-5' high herb, with leaves sheathing the stems and divided into 3 large, toothed and lobed leaflets, each 3-6" broad; white flowers are in large compound umbels, usually appearing flat-topped. Coastal.

NORTHWEST CRIMSON COLUMBINE(see page 41)

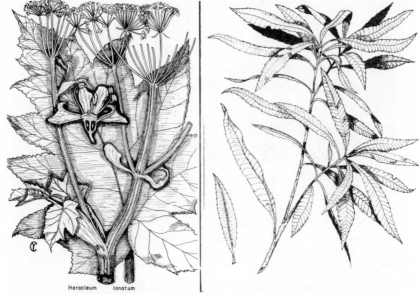

Heracleum   lanatum

WAX MYRTLE (<u>Myrica californica</u>).   One of the typical

Brush Hardw. Conif. plants of the coastal brushfields is this evergreen shrub or tree that grows from 8-30' high.   The nutlet fruit is one of the major foods for band-tailed pigeons at certain times of the year. The thickish, dark green, glossy, oblong or oblanceolate leaves remain on the branches during the winter and make this plant one of the most attractive species found along the coast. Wax myrtles extend from Santa Monica Mts. N. to Wash.

CANYON GOOSEBERRY (see forest section, p. 32).

← CALIFORNIA BLACK-BERRY (<u>Rubus</u> <u>vitifolius</u>). This is a common evergreen bush that trails over the ground or climbs over other vegetation in the coastal brushfields and the

Brush Hardw. Conif.

forest edges. It has slender, straight thorns; doubly serrate leaves that are usually pinnately 3-5 foliolate (3-5 leaflets); that are about 1"across; flowers and oblong berries that are black when ripe. The fruit makes fine jam and jelly, and is an important food for fruit-eating birds and some mammals. It is common from California to British Columbia.

SALMON-BERRY (Rubus spectabilis). This is an erect plant that grows from 6-9' high. The bark is reddish-brown and is sparingly thorned except the sterile shoots, which are very well thorned. The deciduous leaves are formed into three leaflets, which have doubly serrate margins. The flowers are from one to three in a cluster; petals red; berries scarlet to yellow; often eaten by Indians when mixed with other foods; popular with birds.

Brush Conif.

← HIMALAYA BERRY (Rubus thrysanthus). This is an introduced evergreen vine that is common along the coast. It is a robust plant with large, recurved thorns that make it almost impossible for a person to get through thickets formed by this species. The leaves are formed in five leaflets in most cases; the white flowers are relatively large; the berries are dark blackish and used as food by many birds and mammals, especially bears.

Brush

*Brush
Wd.Pr.
Hardw.*

TOYON or CHRIST-
MAS BERRY (Hetero-
meles   arbutifolia)
A large shrub or
small tree with fuzzy
young   branchlets;
leaves 2-4" long, dark
green and shining
above, but pale be-
neath, and toothed;
numerous   small
white flowers in ter-
minal   corymb-like
panicles; fruit be-
comes bright red.
California coast.

VARICOLORED LU-
PINE (Lupinus variicol-
*Brush
Strand* or). The 2-3' long stems usually lie along the ground; the flowers
ers are similar to those of the lupine on page 10, but with flowers
colored variously whitish, yellow, pinkish, purple or bluish.

BLUE-BLOSSOM (Cea-
nothus thyrsiflorus).   In
the spring months along
the roadsides in the red-
wood region and north to
Coos Bay, Oregon, many
Brush brushfields take on a bluish
color from the blossoms of
this 3-25' tall shrub or
small tree. It is an important deer food, sometimes
called blue myrtle or lilac.

Several other kinds of
ceanothus are found in this
region, of which probably
the most important is deer
or buck brush (page 31).

SALAL (see page 34).

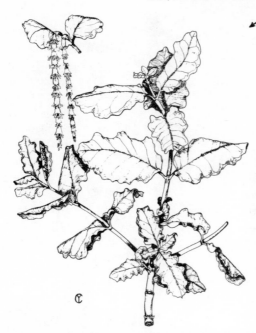

SILK TASSEL (Garrya elliptica). This 5-9' or, more rarely, 20' high shrub or tree has simple, opposite leaves with noticeable undulating margins. It has dioecious flowers (with stamens & pistils on different flowers on different plants),which are borne along a pendulous, catkin-like structure. This plant is common in the coastal ranges on the seaward side from Monterey Co., Calif. to Lane Co., Oregon.

Brush

BEAR BRUSH (Garrya Fremontii). 3-10' high bush with slender branches, smooth bark (the above has rough bark), dark leaves, purplish berries.

Brush

BLACK TWINBERRY (Lonicera involucrata). An erect shrub, 3-10' high, with simple, entire leaves that occur in pairs, as do also the attractive yellow flowers, and the black berries. Below the flowers are large and conspicuously broad bracts, which become reddish as the fruit develops. Most of region.

*Brush
Hardw.
Str.Wd.*

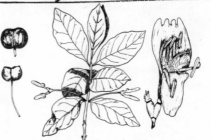

WEDDED HONEYSUCKLE (Lonicera conjugialis). This is a close relative of the twinberry, with blackish-purple flowers, and red berries

*Brush
Str.Wd.*

(at maturity) which are 2/3 joined (as shown). Humboldt Co. & North.

BUSH MONKEY FLOWER (Mimulus aurantiacus). 3-5' bush with large, brownish-yellow, funnel-shaped, narrow-throated flowers. It is easily told by its sticky foliage and dark green leaves. From Curry Co., Oregon, south along the coast.

Mimulus aurantiacus

SEASIDE PAINTED CUP (Castilleia latifolia). 1/2 to 1 1/2' herb with sticky, hairy foliage; the greenish-yellow flowers surrounded by yellow bracts; leaves usually 3-lobed. Sea cliffs.

CASCARA SAGRADA (Rhamnus purshiana). See page 30.

HAIRY MANZANITA (Arctostaphylos columbiana). Page 33.

Sunflower Fam. (tiny flowers concentrated into heads)

LIZARD TAIL (Eriophyllum staechadifolium). A diffusely-spreading, 1 1/2 to 3' high bush with narrow leaves, narrowest at base, and with a dense, felt-like, whitish fuzz underneath, green glabrous above. Flower heads are yellow. From Coos Co. S.

Brush Strand

PEARLY EVERLASTING FLOWER (see page 43).

Brush Strand SEASIDE DAISY (Erigeron glaucus). This beautiful little daisy rises 2-14" high from stems spreading over the ground. It has leaves; flowers that are in solitary, terminal heads or loose corymbs; and large ray flowers, lilac or violet in color. Ore. coast & south.

Brush

COYOTE BRUSH (Bac-charis pilularis). This is a common shrub from 2-5' high that occurs along the coast of Calif. and north to Tillamook Co., Ore. It has small white flower heads & these are either alone or in clusters on the many leafy branchlets. The leaves are obovate or cuneate, entire or few-toothed. The young branchlets are scaly & sticky.

COMMON ANIMALS OF THE COASTAL BRUSHFIELDS

Mammals
Black-tailed deer, 57
Mountain lion, 58
Bobcat, 58
Gray fox, 60
Raccoon, 61
Ring-tailed cat, 61
Long-tailed weasel, 63
Short-tailed weasel, 63
Spotted skunk, 64
Striped skunk, 64
Townsend chipmunk, 67
Mountain beaver, 68
Woodrats, 69
White-footed mouse, 71

Birds
Quail, 83-84
Hummingbirds, 89
Bush-tit, 93

Birds (continued)
Wren-tit, 94
Winter wren, 95
Bewick wren, 94
Pileolated warbler, 97
Scrub jay, 92
American goldfinch, 100
Towhees, 100
Oregon junco, 101
Fox sparrow, 102
Song sparrow, 102

Reptiles & Amphibians
Gopher snake, 109
W. garter snake, 107
W. ring-necked snake, 107
Alligator lizards, 104
Western skink, 104
Western fence lizard, 104
Calif. slender salamander, 113
Eschscholtz's salamander, 113

CONIFEROUS FORESTS

The Pacific coastal coniferous forest is the most characteristic habitat type of this region. It is, in mature form, a dense forest of tall trees (often over 300 feet), of great diameter. The crowns of the trees commonly touch to form an overhead screen which shades out most light from the forest floor. Only the most shade-tolerant plants can survive there. The dominant trees are conifers (cone-bearing trees, with needle-like or flattened, linear leaves, which are evergreen). These trees give a continuous shade to the ground below throughout the year. Where a dominant conifer has died and fallen a variety of shrubs and smaller trees usually spring up in the clearing, to be gradually replaced as a new tall conifer grows up and overshadows them.

The nature of the forest varies within the region, dependent largely upon the amount of rainfall and moisture available. Near the coast, in the more moist sites, the sitka spruce and lowland fir are dominant (particularly in Washington and northern Oregon). Farther inland, in drier sites, giant cedar and western hemlock prevail. On the driest coastal forest sites the Douglas fir dominates the scene. In the southern part of the region, in California, redwoods take the place of the spruce-fir and cedar-hemlock types.

Rain forest, sitka spruce type; courtesy Olympic Nat. Park.

Redwood Grove in Del Norte County, with sword ferns and small herbs dominating undergrowth. Photo reproduced by courtesy of Redwood Empire Association.

Douglas fir forest along Eel River in northwestern California. Photo reproduced by courtesy of Redwood Empire Association.

Interior forest of Washington Coast on Soleduck River, featuring giant cedar and western hemlock. Photo reproduced by permission of National Park Service at Olympic National Park.

The trees described as characteristic of the coastal hardwood forest habitat (see page 45) also occur as understory or second-growth trees in the coniferous forest. Similarly the shrubs of the coastal brushfields (see page 17) often occur as undergrowth or in clearings in the coniferous forests along with other, more shade-tolerant shrubs.

*　　　*

REDWOOD (Sequoia sempervirens). This is a character-istic tree of the belt from southern Oregon to Monterey Co., California. Seldom does this species occur in-land for more than 20 miles. The sharp-pointed, dark green leaves are usually 2-ranked on most of the branches. This tall, massive tree reaches its greatest develop-ment in Humboldt Co., California, where the tallest known tree (367 feet) occurs. Redwoods are important rec-reation and lumber trees.

Conif.

DOUGLAS FIR (Pseudotsuga menziesii). This widely distri-buted tree of the western states and Canada is one of the most im-portant lumber trees of the Pa-cific Northwest, including nor-thern California. The inch-long needles are flattened and usually in 2 ranks on each side of the rib. The 3-forked cone bracts (shown) are different from those of any other conifer, which makes iden-tification easy. Height to 250'.

Conif.

Conif.

SITKA SPRUCE (Picea sitchensis). This large forest tree often grows nearly 200 feet tall in the moist, often swampy situations along the coast. The wood is light brown, with thick, whitish sapwood, light and relatively weak. In the past the wood has been used for shipbuilding and for aeroplanes. It is easily identified by its very stiff, pungent-smelling leaves, which are whitish above and with a ridge in the middle and rounded strongly outward below. Alaska to Mendocino coast.

Conif.

LOWLAND FIR (Abies grandis). The leaves of this attractive tree of the low moist valleys along the coast are 2-ranked and form flat sprays. The upper surface is dark green, the lower silvery-white. The cones are 2.5 to 4" long and stand upright on the branches. The branches are grayish in color. Sonoma Co. N.

Conif.

WESTERN HEMLOCK (Tsuga heterophylla). The light, hard, tough, pale yellow-brown wood of this attractive western tree is important for building purposes, and the bark is used in tanning hides. The weeping-willow-like top is distinctive, as are the grooved and ridged leaves. From the Mendocino coast of California, it is found north along the coast mainly on drier, higher ground than spruce or redwood.

GIANT CEDAR (Thuja plicata). This attractive tree is found in moist bottom land or occasionally on dry ridges from sea level to approximately 6,000' altitude, and from the Mendocino coast north to Alaska. Minute, scale-like leaves thickly cover the branches, which form into flat sprays. The short cones are held on short lateral branchlets. The bark is cinnamon-red and the tree may grow 180' or more high. Formerly Indians burnt out the trunks to make canoes.

Conif.

RED ALDER (Alnus rubra). This fast-growing tree, with light, soft wood that is used for furniture and for smoking salmon in the Pacific Northwest, is abundant along the coast from Santa Barbara Co., California, to Alaska. It grows from 20-90' high and forms groves in the bottom-lands and along creeks. The leaf edges are all double-toothed (serrate) and the bark is pale gray or whitish-mottled.

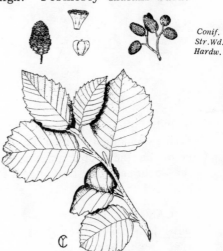

Conif.
Str.Wd.
Hardw.

Quercus   chrysolepis

CANYON or MAUL OAK (Quercus chrysolepis). 30-65' high tree; young twigs covered with whitish fuzz; has smooth, ashy-gray bark; dark green leaves thick and leathery; thick walls to acorn cups covered with rusty fuzz. Canyons from S. Ore. south.

Conif.
Str.Wd.

← WESTERN AZALEA (Rhododendron occidentale). This is one of the most attractive shrubs along the coastal areas in California north into Oregon, with its large, whitish flowers tinged with yellow and pink shades. The blooms appear in May and continue for several weeks; in the fall the deciduous leaves turn bright colors.

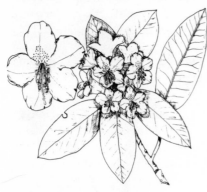

Conif.
Hardw.
Brush

CALIFORNIA RHODODENDRON or ROSE BAY (Rhododendron macrophylum). This is usually an erect shrub from 4-8' high, but in the redwood belt in California it often occurs as a small tree up to 30' high. The large, rose-purple flowers, borne in a cluster, make this one of the primary attractions of the coniferous forests along the west coast in the spring months. The thick, entire-margined, evergreen leaves make this plant attractive also during the winter months. Rhododendrons are found along the California coast and north to Washington and British Columbia.

Conif.
Hardw.
Brush

CASCARA (Rhamnus purshiana). A small tree or shrub from 8-20' high, with thin, deciduous leaves, 3-8" long. The flowers are greenish and the berries black. The bark possesses cathartic properties peculiar to the genus and is collected and sold as

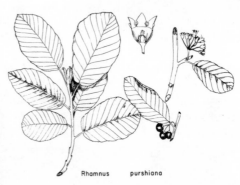

Rhamnus purshiana

Cascara Sagrada. It occurs from central Calif. N. to Wash.

DEER BRUSH (Cea-
nothus integerrimus). A
variety of ceanothus or
wild lilac are found in
California and southern
Oregon. This widely-

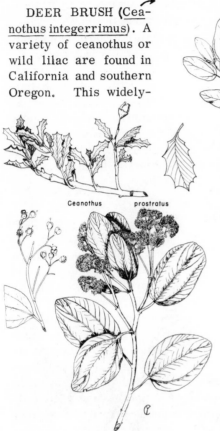

Ceanothus   prostratus

Ceanothus   velutinus

branched shrub is from 4–
12' high and occurs along
the coast or near it as far
north as southern Douglas
Co., Oregon.   The showy
white or blue flowers have
a fragrant smell and are
borne in clusters along a
main branch.   The leaves
are thin, deciduous (falling
off in winter) and entire
margined.   This is an important deer food.

*Conif.
Hardw.
Brush*

SQUAW MAT (Ceanothus
prostratus).   This plant
often groups together   to
form flat patches with the
stems stretching over the
ground.   It is usually found in the pine area of the interior, but
does occur along the coast too.   The toothed, evergreen leaves
are distinctive because of their resemblance to holly.

Conif.

SNOW BUSH or TOBACCO BUSH (Ceanothus velutinus). This
shrub grows up to 12' high; has evergreen, varnished, eliptical
leaves; and large, compound panicles of white flowers.   The
leaves are also prominently 3-nerved. It is a plant of the inland mountains, but occurs also along the coast from Humboldt
Co., California, north to British Columbia.

*Conif.
Brush
Hardw.*

Conif.
Brush
Hardw.

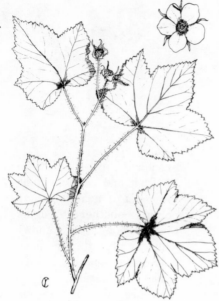

← THIMBLE-BERRY (Ru-bus parviflorus). This common shrub is from 3-6' high with shreddy bark. It has deciduous, simple and palmately-lobed leaves from 3-7" wide, and it is common in the coastal coniferous forests north to Alaska. Its flowers are scarlet, mild in taste and edible.

CANYON GOOSEBERRY (Ribes menziesii). One of the com-

Conif.
Brush
Hardw.

mon gooseberries found along the coast is this well-thorned, 4-8' high shrub. It has white-petaled flowers with red or purple sepals; the berry is a dark purple, about 1/4" in diameter, and covered with stiff, spine-like, gland-tipped bristles. The berries were mixed into other food by the Indians for flavoring.

RED-FLOWERING CURRANT (Ribes sanguineum).
This is one of the first shrubs of the coast to flower. It has blood-red to pinkish flowers, depending on the variety. The unarmed stems are slender, erect or spreading, and are from 3-10' high. Varieties of this species occur from San Luis Obispo Co., California, north to British Columbia. The bluish-black berries are slightly edible.

Conif. Hardw. Brush

PRICKLY CURRANT (R. lacustre). (Not illustrated.) The 3-7' stems usually lie along or near to the ground, with 1-3 spines at each joint; leaves heart-shaped; flowers purple.

Conif. Str. Wd.

HAIRY MANZANITA (Arctostaphylos columbiana). Many manzanitas are found in California, but this common evergreen shrub, covered with hairs and with white flowers borne in clusters, is one of the few that occurs north into Oregon and Washington along the coast. Like all manzanitas, it has a dark, smooth, red bark, and is very leafy, growing from 4-8' high in the immediate coastal area. The berries were ground up

Conif. Rocks Brush

and mixed with other food as flavoring by the Indians.

Conif.
Hardw.
Str.Wd.

BLACK HUCKLEBERRY (Vaccinium ovatum). This attractive evergreen shrub that grows from 4-8' high is one of the dominant cover plants in the redwood region in California. It has shiny, stiff and leathery leaves; and small, pinkish, urn-shaped flowers borne in racemes. The blackish fruits are eaten by many wild animals in the fall; they also make excellent jams and jellies. This species is found along the coast from Santa Barbara County, in California, to British Columbia. Why not try a wild huckleberry pie?

RED HUCKLEBERRY (Vaccinium parvifolium). This shrub grows from 3 to nearly 20' high. The leaves are thin, oval, and deciduous (falling off in winter). The branches and branchlets are greenish and sharply angled; the fruit is bright red and very pleasant to eat; the flower is globular and greenish or pinkish in color. Many birds and animals eat the berries in the fall. This shrub is found from the Santa Cruz Mts., California, north to Alaska.

Conif
Hardw.

SALAL (Gaultheria shallon). (Illustration on top of next page.) This is one of the most common cover shrubs in the redwood region, growing from 1-6' high and often covering the forest floor. The evergreen leaves are shiny above and from 3-6' long. The white to pinkish flowers

Conif.
Hardw.
Brush

are urn-shaped and are borne in racemes. This species is found in the woods from California north to British Columbia. The berries make fine jelly and are eaten by Indians, and by many birds and mammals, especially bears.

OREGON GRAPE (Mahonia nervosa). 2-16' high, evergreen shrub, with pinnately-compound leaves, yellow flowers borne in racemes, dark blue berries and yellow wood. It is found all along the coast. ↓

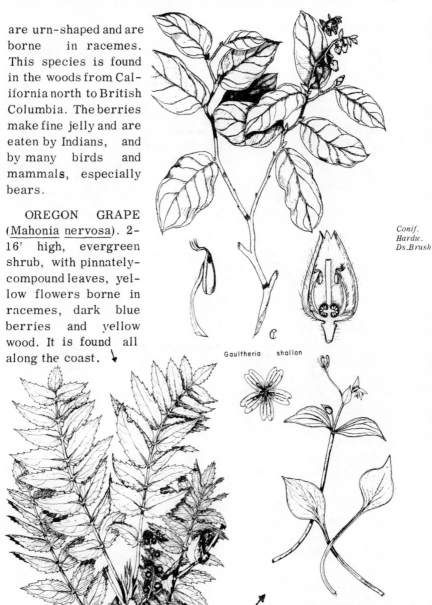

Conif. Hardw. Ds.Brush

Gaultheria shallon

Montia sibirica

Conif. Hardw. Meadow

INDIAN LETTUCE (Montia sibirica). This 9-18" high herb blooms among the earliest flowers in

Mahonia nervosa

moist places in the coastal woods from Calif. N. to Alaska. The flowers are white with pink veins or pink with reddish veins. Edible leaves.

Conif.

REDWOOD SORREL (Oxalis oregana). The obcordate (reversed, heart-shaped) leaves of this small plant that grows in the deep coastal woods are all basal. The single flowers on each scape (flower stem) are pink, white or rose-colored. This sorrel is very common in the redwood forests of the California coast, but less common on north to Washington. It often completely carpets the forest floor.

(The following 12 species belong to the Lily Family, and all have parallel-veined leaves.)

Conif.
Hardw.

FAIRY LANTERN or FAIRY BELL (Disporum smithii). This herb grows about 3' high and is found in the deep woods or along stream banks in the coastal mountains from Santa Cruz Co., California, north to British Columbia. The leaves are sessile and cordate (heart-shaped) at the base. The flowers are creamy-white or greenish, hanging in bell-shaped clusters underneath the leaves so as to be often invisible from above. The fruit is a yellow or reddish berry, which is not edible.

Oaks
Brush
Conif.
Hardw.

COMMON TRILLIUM (Trillium sessile). Not illustrated. It looks like T. ovatum on next page, except the flower petals are sessile, rising directly from the 3 large leaves. 4-12" high.

WESTERN WAKE-ROB-
IN or TRILLIUM (Trillium
ovatum). This is a glabrous,
erect, unbranched herb that
grows from 6-10" high.
There are three bright green
leaves in a whorl at the sum-
mit of the stem. Extending
beyond these leaves is a
white, lily-like flower that
turns to a pinkish-rose col-
or a few days after it has
opened. This plant is common in the woods along the
coast from the Santa Cruz
Mts. of Calif. N. to Br. Col.

*Conif
Hardw.*

BROOK TRILLIUM (T.
rivale). Similar to above
flower, but leaves have separate stems; flowers white.

*Conif
Hardw.
Str. Wd.*

Trillium ovatum

Lilium columbianum

OREGON LILY (Lilium co-
lumbianum). This reddish-
orange lily, with many dark
purple spots, blooms in the
redwood country along the
coast during June and July.
It grows from 2-4' tall and
usually, but not always, has
the leaves in whorls of 3's at
the nodes. It extends from
Humboldt Co., California,
north to British Columbia.

*Str. Wd.
Hardw.
Conif.*

WASHINGTON LILY (L.
Washingtonianum). Has 4-6'
tall stems and 3-3 1/2" long,
white flowers with fine purple
dots, the whole turning purplish with age; leaves 1 1/2 to
5" long, and usually in several whorls. Found in open woods at
middle altitudes from northern Humboldt Co. to N. Oregon.

*Hardw.
Conif.*

See illustration on next page.

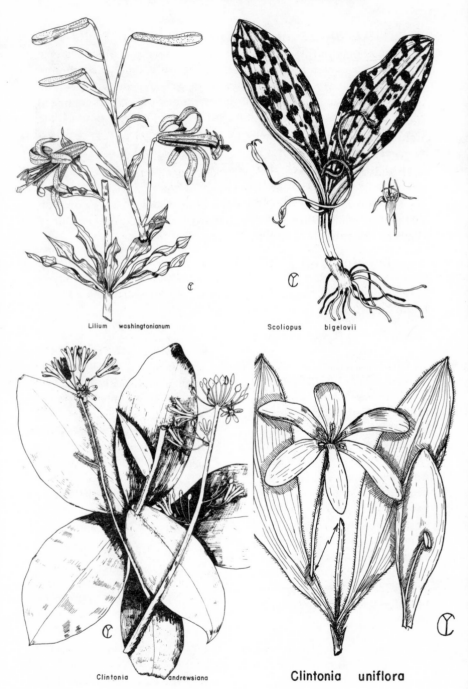

Lilium   washingtonianum

Scoliopus   bigelovii

Clintonia   andrewsiana

Clintonia   uniflora

CALIFORNIA FETID ADDER'S TONGUE or SLINK-POD (Scoliopus bigelovii). This very short, largely subterranean plant produces two large mottled leaves (4-9" long), and an umbel of flowers which are mottled with green and purple. The flowers occur in January through March and the plants appear in deep, cool, shaded areas in the redwood forest from Santa Cruz County north to the southwestern part of Oregon.

RED CLINTONIA (Clintonia andrewsiana). This is a glabrous herb that has large elliptical leaves that are from 7-13" long; the red, lily-like flowers are borne at the top of a long, bare, 15-20" stem. The berries are dark blue. This plant is common in the redwood forests and is found from southwestern Curry County, Oregon, to Monterey County, California.

ONE-FLOWERED CLINTONIA (C. uniflora). Has solitary white flowers on comparatively short stems, and only 2-3 leaves instead of 4-5 (for the above flower). Middle altitudes, Humboldt County north to British Columbia.

OREGON COLTSFOOT or TWO-LEAVED SOLOMON'S SEAL (Maianthemum dilatatum). This glabrous (smooth) herb that grows on moist shaded banks in the coastal woods from Marin County, California, north along the coast to Alaska, usually has two large, cordate (heart-shaped) leaves. Above these it bears a terminal raceme of small white flowers. The stem is simple and erect and grows from a few inches to 14" high. A bright red berry.

Conif.
Hardw.

*Conif.
Hardw.
Oaks*

FALSE SOLOMON'S SEAL

(Smilacina racemosa).   A 1-4' high
herb with two flat rows of oblong to
oblong-lanceolate leaves; usually
sessile and with tiny hairs beneath;
hairy stems topped by a dense panicle
of tiny white flowers, which turn into
round, red-mottled berries.   From
Monterey C. north to Br. Col.

Smilacina racemosa

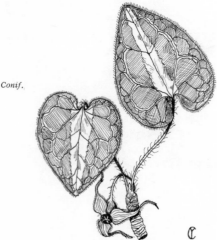

Conif.

Asarum   hartwegi

WILD GINGER (Asarum
caudatum).   This perennial
herb, with aromatic, creeping rootstocks, has basal,
cordate   (heart-shaped)
leaves that are dark green
above and lighter below.   It
has a conspicuous, solitary,
large, brownish flower that
is borne from a lower axil
(point of leaf attachment)
near the ground.   It is a
plant of the moist, shaded
woods from British Colum-
bia S. to Sta. Cruz Mts., Cal.

MARBLED WILD GIN-

Conif.   GER (Asarum Hartwegi). Very similar, but leaves are marbled.

NORTHWEST CRIMSON COLUMBINE (Aquilegia formosa). This plant has red spurs and sepals and yellow petals. It is found mainly inland in open woods, but does occur in the coastal mountains and along the coast from southern California north to British Columbia.

VAN HOUTTE'S COLUMBINE (A. eximia). Similar to the above, but sticky, hairy all over. Mendocino Co. to San Mateo Co., California.

Conif.
Hardw.
Brush
Rocks

Rocks
Conif.
Hardw.
Brush

INSIDE-OUT FLOWER (Vancouveria hexandra). This low, creeping, perennial herb grows from slender, creeping rootstocks. The white, nodding flowers are reflexed (which means the petals and sepals point in an opposite direction from the stamens) and this gives the plant its name. They are borne on an open panicle at the top of an 8-21" high stem. This species is found in the open woods from Mendocino Co., California N. to Washington.

Conif.
Hardw.
Oak

SMALL-FLOWERED VANCOUVERIA (V. planipetala). Flowers more numerous than in above, and covered with sticky hairs (whereas V. hexandra is smooth). Redwood forests & N. to Curry Co., Or.

Conif.
Oak
Hardw.

Conif.
Hardw.

DEER-FOOT (Achlys tri-phylla). This perennial herb, which is about 1' high, spreads by slender root-stocks. The leaves consist of three sessile (without stems) leaflets about 2-6" across, while the flowers have no se-pals or petals and are borne on a short, dense spike. The fruits are dry, moon-shaped seeds. Deer-foot is common in the coastal woods from Mendocino County, Califor-nia, north through Oregon and Washington to British Columbia.

Conif.
Hardw.

Conif.
Hardw.

Dicentra   formosa

← PACIFIC   BLEEDING HEART (Dicentra formosa). The beautiful rose-purple flowers of this plant top an 8-18" stem that is surroun-ded by basal leaves, each leaf being on a long stem and finely-divided into separate parts. It blooms in the spring from the San Fran-cisco Bay Region north to British Columbia.

ONE-FLOWERED DI-CENTRA (D. uniflora). The small, white or pinkish flow-er (sometimes there are 2) tops a 1-3" stem, surround-ed by a few much-divided leaves. The outer petals are more strongly recurved than in formosa. Humboldt Co. N. to Wash.

(See illustration on next page).

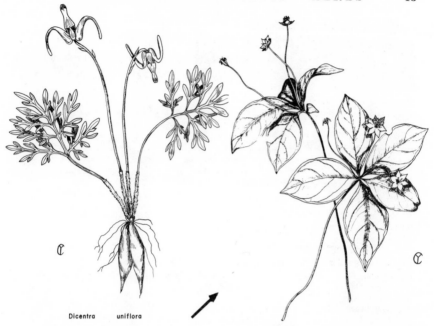

Dicentra uniflora

STAR FLOWER (<u>Trientalis europaea</u>). This 2-6" high, glabrous plant is common in the woods along the coast from Monterey County, California, north to Alaska. The star-like white to pinkish flowers grow on thin stalks that spring from a whorl of 3 to 6 leaves that are usually from 1-2" long.

Conif. Oak Hardw.

PEARLY EVERLASTING (<u>Anaphalis margaritacea</u>). This 1-2' high composite (with flowers combined into dense heads) blooms in the fall and is often used for winter bouquets. The yellow flowers are surrounded by pearly-white bracts and the heads are arranged in compound corymbs. The bracts form overlapping series, which are persistent long after the flowers have dried up. From Monterey County north to Alaska.

Conif. Brush Hardw.

Anaphalis margaritacea

Hardw.
Conif.
Str.Wd.

LOWLAND   CUDWEED.

(Gnaphalium palustre). 3-9" high herb, branching at the base, the stems and leaves covered with long, loose wool (often falling off the leaves), in which the flowers are imbedded. Low places.

COMMON ANIMALS OF THE CONIFEROUS FORESTS

Mammals
Roosevelt elk, 56
Black-tailed deer, 57
Black bear, 58
Bobcat, 59
Ring-tailed cat, 61
Marten, 62
Fisher, 62
Weasels, 63
Skunks, 64
Snowshoe hare, 65
Chickaree or red
   squirrel, 66
Flying squirrel, 66
Townsend chipmunk 67
Mountain beaver 68
Wood rat, 69
White-footed mouse, 71
Red tree mouse, 71
Calif. red-backed vole, 71
Townsend mole 73
Shrew-mole 73
Pacific water shrew, 73
Trowbridge shrew, 73
Pacific shrew 73
Dusky shrew, 73
Bats, 74

Birds
Sharp-shinned hawk, 81
Blue grouse, 83
Horned owl, 88
Screech owl, 87
Spotted owl, 88
Pacific nighthawk, 88
Vaux swift, 89
Allen hummingbird, 89
Yellow-bellied sapsucker, 90
Red-shafted flicker, 90
Pileated woodpecker, 90
Hairy woodpecker, 90
Downy woodpecker, 90
Olive-sided flycatcher, 91

Birds (continued)
Western wood peewee, 91
Western flycatcher, 91
Violet-green swallow, 92
Tree swallow, 92
Purple martin, 92
Canada jay, 92
Steller jay, 93
American crow, 93
Chestnut-backed chicka-
   dee, 93
Red-breasted nuthatch, 94
Pygmy nuthatch, 94
Brown creeper, 94
Robin, 95
Varied thrush, 96
Russet-backed thrush, 96
Hermit thrush, 96
Mexican or Western
   bluebird, 95
Winter wren, 95
House wren, 95
Golden-crowned kinglet, 96
Ruby-crowned kinglet, 96
Cedar waxwing, 97
Hutton vireo, 97
Townsend warbler, 98
Orange-crowned warbler, 97
Audubon warbler, 98
Black-throated gray
   warbler, 98
Pileolated warbler, 97
Myrtle warbler, 98
Western tanager, 99
Black-headed grosbeak, 99
Evening grosbeak, 99
Purple finch, 99
Pine siskin, 100
Red crossbill, 100
Spotted towhee, 100
Oregon junco, 101
Fox sparrow, 102
Chipping sparrow, 101

Reptiles
Western fence lizard, 104
Western skink, 104
No. alligator lizard, 104
Rubber snake, 108
Western ring-necked
   snake, 107
Sharp-tailed snake, 107
Gopher snake, 109
Northwestern garter
   snake, 107
Western garter snake, 107
Common garter snake, 109

Amphibians
Olympic salamander, 110
Pacific giant salamander, 114
Northwestern salamander,
   109
Long-toed salamander, 109
Rough-skinned newt, 114
Western red-bellied
   newt, 110
Dunn's salamander, 110
Del Norte salamander, 110
Western red-backed
   salamander, 113
Van Dyke's salamander, 113
Eschscholts's salamander,
   113
California slender
   salamander, 113
Clouded salamander, 113
Black salamander, 113
Western toad, 114
Tailed frog, 113
Pacific tree-frog, 114
Yellow-legged frog, 114
Red-legged frog, 114

COASTAL HARDWOOD FOREST

Hardwood forests as such are nowhere widespread within this region. However, they are frequently interspersed among the coniferous forests, as hardwood stands occupy sites that, for one reason or another, are not suited to the taller, coniferous trees. Hardwood stands also frequently replace the coniferous forests following fire, logging, or other disturbance. Once established, they may maintain themselves for many years before finally being replaced by conifers.

Most of the hardwood forests of this region are made up of evergreen, broad-leaved trees. The madrone and tan oak are the two most characteristic species. In addition to these broadleaved evergreen species, such deciduous, broad-leaved trees as the big-leaf maple and the red alder (see page 29) often form groves particularly in the more moist sites.

The understory vegetation of the hardwood forests differs little from that of the coniferous forests, although the hardwood trees themselves provide a source of food for species of animals that are not so well adapted to the coniferous groves.

Mixed hardwood growing near Maple Creek, Humboldt County. with madrone, bay, alder & oak. Photo by William McKittrick.

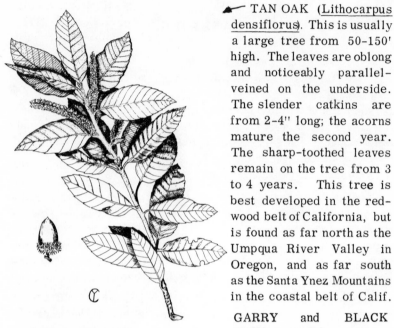

Hardw.
Conif.

TAN OAK (<u>Lithocarpus densiflorus</u>). This is usually a large tree from 50-150' high. The leaves are oblong and noticeably parallel-veined on the underside. The slender catkins are from 2-4" long; the acorns mature the second year. The sharp-toothed leaves remain on the tree from 3 to 4 years. This tree is best developed in the red-wood belt of California, but is found as far north as the Umpqua River Valley in Oregon, and as far south as the Santa Ynez Mountains in the coastal belt of Calif.

GARRY and BLACK OAKS (page 51).

CANYON OAK (See page 29)

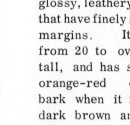

MADRONE (<u>Arbutus menziesii</u>).

Hardw.
Conif.

This is an evergreen tree with glossy, leathery leaves that have finely serrate margins. It grows from 20 to over 100' tall, and has smooth, orange-red colored bark when it is new, dark brown and fis-sured into small scales when it is old. Flowers are white; fruits are red or orange berries. It is common from Monterey Co., Calif. north to British Columbia.

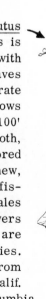

CALIFORNIA LAUR-EL or OREGON MYR-TLE (Umbellularia californica). This is a dense-crowned, evergreen tree that grows from 20-60' high. It is usually found close to water, and is especially common in the alluvial flood plains along the rivers in northwestern California and southwestern Oregon, but is also found on shady hillsides and canyons. The fine wood is used in cabinet making and souvenirs. It is found as far south as San Diego Co. in California.

Hardw. Str.Wd.

BIG-LEAF MAPLE

(Acer macrophyllum). This broad-crowned tree grows from 30-100' tall, and has large five-lobed leaves that are from 4-10" broad. The small, greenish-yellow, bell-shaped flowers occur in corymbs or dense, cylindrical racemes. The winged fruits are hairy. This species occurs west of the crest of the Sierra Nevada throughout California and north along the west coast to Alaska.

Hardw. Str.Wd.

Hardw.
Conif.
Brush

Hardw.
Brush

Castanopsis sempervirens

ℭ

← BUSH CHINQUAPIN
(Castanopsis sempervi-
rens).   This evergreen,
spreading shrub is from
2-8' high and has smooth
brownish bark. The ob-
tuse leaves are usually
lanceolate-oblong and en-
tire.   The catkins are
usually branched and the
nutlets are borne   in
chestnut-like burs. This
species is found in the
coastal mountains of Cal-
ifornia and southern Ore-
gon, but it is not   found
near the sea.

CHINQUAPIN (C. chry-
sophylla). A tree, 40–90'
high with   rusty, scaly
branchlets and thick bark;
leaves   sharp-pointed.
Sometimes a 3-15' shrub
with clustered branches.
Washington to Calif.

FIRE-WEED (Epilobium
angustifolium) This perennial herb grows from 2-6'
high; the leaves are entire,
narrow and 4-6'' long. The
flowers are lilac-purple and
often appear   flame-like
among the greenery. Fire-
weed is common in   the
moist, burnt-over areas in
the north Coast Range of
Calif. north to Alaska.

Hardw.
Brush
Str.Wd.

NORTHERN WILLOW-
HERB (E. adenocaulon). 1–
3 1/2' tall; flowers & flower stems sticky-hairy; petals white.

Hardw.
Brush

ℭ

Epilobium angustifolium

COMMON ASTER (Aster chilensis). A common fall-blooming, white, bluish-red or bluish aster of the hills along the coast of N. Calif., growing 4-16" high; leaves usually hairy and the margins scaly.

Hardw.
Meadow
Brush
Wd.Pr.

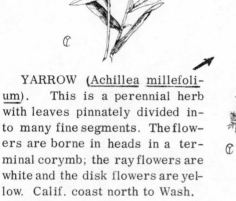

YARROW (Achillea millefolium). This is a perennial herb with leaves pinnately divided into many fine segments. The flowers are borne in heads in a terminal corymb; the ray flowers are white and the disk flowers are yellow. Calif. coast north to Wash.

Hardw.
Brush

COMMON ANIMALS OF THE COASTAL HARDWOOD FOREST

Mammals
Most of the species found in the coniferous forests (page 44) also occur in this habitat, but particularly the following:

Chickaree, 66
Wood rat, 69
White-footed mouse, 71

Birds
Many coniferous forest species (page 44) are found here, but particularly the following:

Sharp-shinned hawk, 81
Blue grouse, 83

Birds (continued)
Ruffed grouse, 83
Band-tailed pigeon, 87
Belted kingfisher, 89
  (along streams)
Downy woodpecker, 90
Yellow-bellied
  sapsucker, 90
Flycatchers, 91
Bush-tit, 93
Dipper (by streams), 95
Varied thrush, 96
Hermit thrush, 96
Russet-backed thrush, 96
Kinglets, 96
Cedar waxwing, 97
Hutton vireo, 97
Warblers, 97-98

Evening grosbeak, 99
Oregon junco, 101
Fox sparrow, 102

Reptiles
Western fence lizard, 104
Western skink, 104
No. alligator lizard, 108
Foothill alligator lizard, 104
Rubber snake, 108
W. ring-necked snake, 107
Sharp-tailed snake, 107
Gopher snake, 109
Garter snakes, 107, 109

Amphibians
Most amphibians are found in this habitat, pages 109-115, except bull and spotted frogs.

WOODLAND-PRAIRIE

Throughout the coastal forest region grasslands occur. Some of these are natural, reflecting a difference in soils. Others are man–caused, the result of burning tree and shrub cover, and seeding the burned-over lands to grasses. In either case, the grasslands, known locally as prairies, support a different group of plants and a characteristic animal population.

Woodland-Prairie. Photo from files of U. S. Forest Service.

In the interior of the region, in the coastal mountains, are many areas in which the vegetation represents an intermixture of woodland and prairie. The woodland usually consists of stands of Garry Oak, sometimes with Black Oak in combination, and a variety of understory plants similar to those described in this book under Hardwood Forest (page 45) and Coastal Brushfields (p. 17). The oak stands vary from dense, close-canopy forests to an open savanna with scattered trees in a predominantly grassland area. The trees are all deciduous hardwoods, shedding their leaves in winter. They produce an abundance of acorns to provide food for a variety of animals. The grasses are of numerous kinds, of which three very common ones are described here.

Wd.Pr.
Oak

GARRY or OREGON WHITE OAK (Quercus garryana). This is a typical, round-topped oak tree that grows from 25-60′ tall. The lobed leaves are from 3-6″ long; the acorns are smooth and shiny and are important deer food in the fall of the year, while the Indians ground them up for acorn mush and bread. This oak is found all along the coast and inland to the base of the Cascades from British Columbia south to the Santa Cruz Mts. in California.

CALIFORNIA BLACK OAK (Q. kelloggii). This broad, round-topped tree grows from 30-80′ high. The leaves are 3-lobed on each side and have from one to three bristle-tipped teeth. The acorns are deep-set in the cup. This oak occurs in the Coast Ranges as far north as southwestern Oregon, but does not occur near the sea. The acorns were ground into meal and leached of their poisonous tannin by pouring boiling water over the meal in a sand pit when the Indians prepared them for eating.

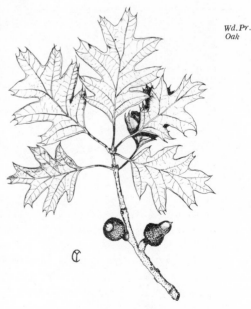

Wd.Pr.
Oak

CALIFORNIA BUCKEYE (Aesculus californica).

A shrubby tree, seldom over 30' high with a short smooth trunk and a flat open topped crown. The bark is light grey. Dark green leaflets 3-7" long are palmately arranged in groups of five on a petiole 4-5" in length. The flowers are pinkish white and grow in erect panicles. The fruit is pear shaped containing one seed that is about 2" in diameter.

Aesculus   californica

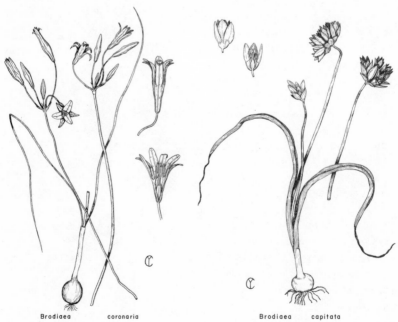

Brodiaea   coronaria   Brodiaea   capitata

WESTERN WILD OAT-GRASS (Danthonia californica). 2-3' high, tufted perennial with spreading branches and usually purplish bracts surrounding the flowers. North to British Columbia. *Wd.Pr.*

PACIFIC REED GRASS (Calamagrostis nutkaensis). 2 1/2 to 5' high, with very long stiff leaves, and a narrow panicle of pale green or purplish spikelets. From California to Alaska. *Wd.Pr.*

VELVET GRASS (Holcus lanatus). 1-2' high grass with very distinctive covering of tiny, velvety hairs; narrow panicles of purple-tinged spikelets. Coast Ranges north to Washington. *Wd.Pr.*

← FIRE-CRACKER PLANT (Brodiaea ida-maia). The red flowers tipped with yellow are borne on a stalk that grows from 1-3' high. This attractive plant grows near Oregon Caves, Oregon, and south through the Trinity Mountains to Marin and Contra Costa counties, California. It is a plant of the hillsides in open areas and is not found along the ocean coast. *Wd.Pr.*

HARVEST BRODIAEA (B. coronaria). Violet-purple flowers with long, droopy petals. *Wd.Pr.*

COMMON BRODIAEA (B. capitata). Flower stem twisting, higher than leaves, and topped by dense head of deep, violet purple (or rarely) reddish-purple flowers. All the stamens have yellow anthers on top.

COMMON MONKEY FLOWER (Mimulus guttatus). 1-3' high herb, found in moist areas from California N. to Alaska. The flowers are

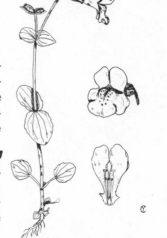

Wd.Pr.

Wd.Pr. Meadow Str.Wd.

yellow with brown or purple dots in the flower throat, and are usually found in terminal racemes.

_____

COMMON ANIMALS OF THE WOODLAND-PRAIRIE

Mammals
Black-tailed deer, 57
Black bear, 58
Mountain lion, 58
Bobcat, 59
Coyote, 59
Ring-tailed cat, 61
Badger, 64
Striped skunk, 64
Black-tailed jackrabbit, 65
Gray squirrel, 66
Chickaree, 66
Ground squirrel, 67
Pocket gopher, 67
Meadow vole, 71
White-footed mouse, 71
Moles, 73
Bats, 75

Birds
Turkey vulture, 81
Hawks, 81-82
Blue grouse, 83
Ruffed grouse, 83
Mountain quail, 83
California quail, 84
Band-tailed pigeon, 87
Mourning dove, 87
Most owls, 87-88
Pacific nighthawk, 88
Downy woodpecker, 90
Yellow-bellied sapsucker, 90
Red-shafted flicker, 90

Birds (continued)
Western wood pewee, 91
Raven, 93
Crow, 93
Bush-tit, 93
Mexican bluebird, 95
Kinglets, 96
Warblers, 98
Western tanager, 99
Black-headed grosbeak, 99
Purple finch, 99
Goldfinches, 100
Oregon junco, 101
Chipping sparrow, 101
Fox sparrow, 102

Reptiles
Western fence lizard, 104
Western skink, 104
Foothill alligator lizard, 104
Racer, 107
Western ring-necked snake, 107
Common king snake, 107
Gopher snake, 109
Western garter snake, 107
Common garter snake, 109
Western rattlesnake, 104

Amphibians
Rough-skinned newt, 114
Western red-bellied newt, 110
Eschscholtz's salamander, 113
Calif. slender salamander, 113
Western toad, 114

COMMON ANIMALS
MAMMALS

Mammals are those animals which have hair and give milk to their young. Most people call mammals simply "animals", but scientists apply this name to all living things except plants and viruses. Mammals are among the most complex of living things. The study of mammals is a profitable field for the amateur and systematic observation, combined with careful note-taking, can reveal new and important facts about many species.

Most mammals are not seen by most people, even when they live in large numbers near the edges of towns. This is partly because of their inconspicuous ways, their camouflaging coloring, and their nocturnal habits. The times when most mammals will be seen are early morning and late evening. Motorists will sometimes encounter them on roads at night, picked up briefly by the headlights of the cars.

The larger mammals are easy to identify, the smaller ones much more difficult. As a guide to identification we have compared them in the following pages to the size of four common and well-known mammals: House Mouse (3-4")

House Rat (8-11")

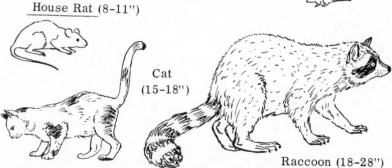

Cat (15-18")

Raccoon (18-28")

In the mammal descriptions mouse + means slightly larger than a house mouse rat - means slightly smaller than a house rat, and so forth. Only the length of head and body, not the tail, is stated and compared.

By comparing sizes, looking at the pictures, reading the descriptions, and noting the type of habitat in which the mammal usually occurs, you will be able to name most of the species which you will encounter.

HOOFED MAMMALS
(All of Deer Family or Cervidae)

Meadow
Conif.
Hardw.

ROOSEVELT ELK (<u>Cervus</u> <u>canadensis</u>). Height 4 1/2 to 5'. This is one of the largest deer, with individuals weighing between 500 and 900 pounds. Antlers are massive, with a single main beam from which the tines (or points) branch forward, and are carried only by males. The antlers are shed in winter and re-grown again in the spring. During the growth period they are covered with velvety skin which is shed before the fall rutting season.

Elk were originally found throughout the region, but are now common only in coastal Oregon and Washington. In California a few herds remain at Prairie Creek and Maple Creek in Humboldt County and in the high mountains of Del Norte County. The Prairie Creek herd is usually near Highway 101 and may be seen by the auto traveller.

Elk are at their best in the fall rutting season, at which time the bulls form harems of cows and sometimes fight spectacular battles with rival males. Their principal habitat is dense forest and brush-fields particularly where interspersed with meadows or prairies.

COLUMBIAN BLACK-TAILED DEER (Odocoileus hemionus). *Meadow* *Wd.Pr.* Height 3 to 3 1/2'. Tail black or deep brown on top, white under- *Brush* neath. The antlers of the adult males are usually forked, and, *Hardw.* *Conif.* in older males, each tine often forks so that four or more points may occur. This is the only common small deer in the region.

This deer is common throughout the region and is often seen feeding in open glades or low brushfields, or in the woodland-prairie region. It is rare in dense forests. The breeding season in this species takes place in the fall and early winter, at which time the bucks lose much of their normal caution and may often be seen in prominent, open places. The small, spotted fawns are born in spring and early summer. Where food is abundant most does will have twin fawns. Sometimes as many as 80 to 100 of these deer have been recorded per square mile.

Formerly the WHITE-TAILED DEER (O. virginianus) was a common animal in central Oregon and Washington, but is now restricted to remnant herds near Roseburg, Oregon, and the mouth of the Columbia River in Washington. This deer is easily distinguished by its long, light brown tail, and by the antlers of the males, in which the tines are directed backward from a single main beam.

ORDER OF CARNIVORES (CARNIVORA)

*Brush
Wd.Pr.
Hardw.
Conif.*

THE   BEAR
FAM.   (Ursidae).

BLACK BEAR
(Ursus american-
us).   Height 2-3'.
With the extinc-
tion of the grizzly
bear, the only re-
maining   member
of the bear family
in the coast region
is the black bear.
This bear ranges in size from 200-500 pounds. The color varies
from cinnamon brown to black, with most being black. Although

sometimes dangerously
friendly in the national
parks, the bear is gen-
erally a wary animal,
rarely seen, while com-
mon from Sonoma Co. N.

CAT FAM. (Felidae)

MOUNTAIN   LION
(Felis concolor). Length
3 1/2 to 4 1/2'.   Otherwise known as the cou-
gar or puma, this is the
largest member of the cat
family in our region, and the
only one with a long tail. The
color ranges from tawny to
grayish, with the tip of the tail
dark brown.   Although widely
distributed in the region, it is
extremely shy and is rarely
seen. It feeds mainly on deer
and is a healthful preventer of
overpopulation.

*Brush
Wd.Pr.
Rocks*

Bobcat

BOBCAT (Lynx rufus). See illustration on page 54. Raccoon size. The bobcat or wildcat is a common animal in the region, but its wary disposition prevents it from being often seen. It is nevertheless more likely to be abroad in the daytime than any other large carnivore. Twice the size of a large house cat, it has a short, bobbed tail and tufted ears. It is often found stalking softly after such small mammals as the brush rabbit and woodrat, or quietly lying in wait for them along a trail, but a large bobcat can attack and kill a deer. Like other carnivores, the bobcat is an important natural check on numbers of smaller mammals and should be protected except in areas where it is actually known to be doing damage.

Strand Wd.Pr.

THE DOG FAMILY (Canidae)

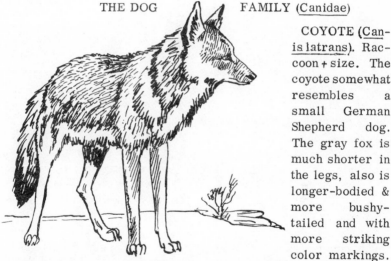

COYOTE (Canis latrans). Raccoon + size. The coyote somewhat resembles a small German Shepherd dog. The gray fox is much shorter in the legs, also is longer-bodied & more bushy-tailed and with more striking color markings.

Most habitats

The coyote is occasionally seen in the wild as it trots across openings, or hunts mice in meadows. It is more often heard than seen. A series of high-pitched yaps, howls and barks constitute the coyote "song", heard most often at dusk or in early morning. It is a relative newcomer to this area, however, having become common only since the opening up of the country by the fires and logging activities of the white men. Where livestock do not occur the coyote is a colorful and useful member of the native fauna and should be protected. Usually, when a coyote runs, the tail is held between the legs, whereas a dog generally carries its tail higher when running.

Most habitats   GRAY FOX (<u>Urocyon</u> <u>cinerioargenteus</u>). Raccoon size. This fox is much less dog-like than the coyote and, in many of its postures and actions, resembles a cat. It is a dainty, neat-looking animal with conspicuous, contrasting gray, black and reddish-brown markings. The long, bushy tail, held straight out when the animal is running, is distinctive. Although both coyote and fox usually avoid heavy timber, the fox is found more commonly in brushy areas than is the coyote, and may be seen along road or trail sides or hunting mice in grassy fields. The fox is much less noisy than the coyote, and its call usually consists of one or a few subdued barks or yaps. It is found as far north as southern Oregon. The RED FOX (<u>Vulpes</u> <u>fulva</u>) occurs in the coastal region of northern Oregon and Washington. It may readily be distinguished by its usually yellowish to orange-red color, and very large and bushy tail. It is more typical of the high mountains.

THE RACCOON FAMILY (<u>Procyonidae</u>)

RACCOON (Procyon lotor). See illustration on page 56. The black, robber mask and bushy, ringed tail serve to distinguish the raccoon from most other mammals.   While the ring-tailed cat has similar tail-markings, it is a much more slender animal, lacks the face mask, and has a much longer tail than the raccoon.   Where the raccoon gives the appearance of a short, bulky, relatively slow-moving animal, the ring-tailed cat is lithe, slender and much more rapid and graceful in movement.   The raccoon is frequently seen crossing or travelling down roads or highways at night, and is a common highway casualty throughout the region.   It likes wooded areas near water, and its baby-like, flat-heeled tracks are common in mud and sand.

Str.Wd.
Strand
Wd.Pr.
Hardw.

CACOMISTLE FAMILY (Bassariscidae)

RING-TAILED CAT (Bassariscus astutus).   House cat size with long, ringed tail, large ears and eyes.   It is primarily an animal of warmer, drier regions to the south and west, but is fairly numerous in our region in northern California and southern Oregon.   It is almost entirely nocturnal and is rarely seen, even in areas where it is relatively common.   It is a quiet hunter of small rodents, which it stalks in the narrow, brushy trails.

Brush

THE WEASEL FAMILY
(Mustelidae)

Conif.

MARTEN (Martes americana). House cat -. This slender, graceful animal has a bushy, somewhat squirrel-like tail, cat-like ears and fur that is golden-brown in color. Probably Olympic National Park in Washington has the greatest number of martens of any place in the region. Normally it is an animal of the higher mountains, but it is also present in the densely-forested country along the coast. It hunts squirrels in the trees.

Conif.

FISHER (Martes pennanti). House cat +. Even more rare and less likely to be seen than the marten is the fisher. Like the marten it probably reaches greatest abundance in the Olympic Peninsula, but is found occasionally in the wilder areas. The fisher is larger than the marten, and is darker in color, ranging from brownish-gray to almost black. Like the marten it feeds mainly on squirrels.

Water
Marsh
Conif.
Hardw.
Strand

MINK (Mustela vison). House cat -. The mink is about the size of the marten, but has much smaller ears and is dark-brown in color. Unlike the marten and fisher, the mink is fairly common in the coastal region. It is largely confined to the vicinity of streams, lake and ocean margins. It is an expert swimmer.

RIVER OTTER (Lutra canadensis). Raccoon size. The otter somewhat resembles a small seal, but with legs and long tail. The tail, thick at the base and tapering to a point, is used as a rudder in swimming. The fur is short and soft; the ears short and inconspicuous; and the feet webbed. All of these help in swimming and the catching of fish, at which it is expert. Otters are playful animals and often make clay slides on the banks for splashing into the water. Found throughout region.

Water Marsh Str.Wd.

LONG-TAILED WEASEL (Mustela frenata). Rat size. This is the most common of the weasel-like animals, and is larger than the short-tailed weasel, as well as lacking the white line which runs down the inside of the hind leg in the latter species. It is generally brown in color, but with whitish on forehead and belly, though the whole body may change to white in winter if there is heavy snowfall. It does not do so in most of the coastal region. It is not rat-like in appearance, and has a 5-6" long tail. When moving it has an undulating, almost snake-like appearance. It often stands with back arched. It twists and turns very swiftly as it follows down a mouse or rabbit.

Brush Wd.-Pr. Hardw. Rocks

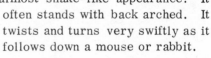

SHORT-TAILED WEASEL (Mustela erminea). Rat -. Much smaller and shorter-tailed that the long-tailed weasel and with a white line on the inside of the hind leg. It is also known as the ermine and becomes white in snowy areas in winter; usually brown with white belly.

Most habitats

BADGER (<u>Taxidea</u> <u>taxus</u>).   Raccoon size.   The short legs, heavy, wide body, and distinctive black and white markings on the face make the badger difficult to mistake for any other animal.

Its strongly-developed, digging claws and feet, which make efficient shovels, help it dig rapidly into the ground after burrowing animals, such as ground squirrels, on which it feeds.   It occurs as far north as Eureka on the coast.

SPOTTED SKUNK (<u>Spilogale</u> <u>putorius</u>).   Rat size.   The broken, white lines and the white-tipped tail easily distinguish this skunk from the larger striped skunk.   It feeds on insects, small animals & plants.

STRIPED SKUNK (<u>Meph-itis</u> <u>mephitis</u>).   Cat size. The two broad white stripes down the back are distinctive.   Both skunks stamp their feet and spread their tail hairs stiffly when alarmed, and throw a very strong-smelling and irritating scent when attacked. This species also eats omnivorously.   It is more commonly seen on roads at night and in more open country.

HARE AND PIKA ORDER (LAGOMORPHA)

HARE FAM. (Leporidae)

BLACK-TAILED JACK-RABBIT (Lepus californicus). Cat + size.

This is an animal of the open country which enters the coastal area only where natural or man-made clearings are found. It extends northward into Oregon, but is found there only in inland valleys. It does not use burrows.

Wd.Pr.
Strand
Op.Brush
Meadow

SNOWSHOE HARE (Lepus americanus). Cat size.

Brush
Hardw.
Conif.

The smaller size, shorter legs and ears and more compact body distinguish this animal from the black-tailed jackrabbit. Its larger size, and darker, more rusty brown color serve to distinguish it from the brush rabbit. From south Oregon north. It usually turns white if much snow.

BRUSH RABBIT (Sylvilagus bachmani). Rat + size.

Unlike the two hares, the brush rabbit is primarily an animal of the coastal forest and brush. It differs from the cottontail of inland regions by its darker color, less white on the tail, and its shorter ears. It does not extend north of the Columbia River, and is always found close to dense cover.

Brush

RODENTS (RODENTIA)

SQUIRREL FAM. (Sciuridae)

Oak
Hardw.
Conif.

WESTERN GRAY SQUIR-
REL (Sciurus griseus).   A
rat + size, all gray squirrel,
with a long, bushy tail and
conspicuous white under-
parts. The Calif. ground
squirrel lacks the bushy tail
and has dark shoulders. The
chickaree is darker and is
more reddish.   The gray
squirrel lives mostly inland.

Conif
Hardw.
Oak

CHICKAREE (Tamiascurus
douglasii). Rat size.   The
chickaree or Douglas Squirrel
is a small, reddish, highly active and noisy tree dweller,
feeding on the seeds of douglas
fir and other conifers.   Its
presence may be detected by
finding piles of remnant fir
cones dropped below feeding
stations. It is found all through
the region. Feeds on nuts.

Conif.

NORTHERN FLYING SQUIR-
REL (Glaucomys sabrinus). It
is rat - The conspicuous
flap of skin that connects the
wrist and ankle of this animal,
and serves as a gliding wing,
distinguishes it from any other.
The large eyes are proof of its
nocturnal nature, for it is rarely active except at night.   It is
chestnut-brown in color.   It can
glide 50 yards or more from a
branch or trunk to a lower one.
Feeds on eggs, insects & nuts.

TOWNSEND CHIPMUNK (Eutamias townsendi). Rat -. This characteristic animal of the coastal forest is larger and darker in color than most other chipmunks. Its stripes on the head distinguish it from the somewhat similar Golden-mantled Ground Squirrel (Citellus lateralis), which may enter the region in the drier, interior mt. areas. The similar-appearing SONOMA CHIPMUNK (Eutamias sonomae) occurs from Del Norte Co. south to the San Francisco Bay. It is smaller and tends to inhabit drier areas. Both chipmunks do little tree climbing, living mainly on ground.

Conif.
Brush
Hardw.
Rocks

Op.Brush
Rocks

CALIFORNIA GROUND SQUIRREL (Citellus beecheyi). Rat size. Otherwise known as Douglas ground squirrel or Beechey, this animal is familiar in the drier grasslands of California and Oregon, and is, therefore, not typical of our region. It is occasionally seen climbing oak trees, but quickly runs down when frightened and not up, as does the gray squirrel. It also has a chunkier body and a browner coat, with dark shoulder colors, and a shorter, less bushy tail. Most often it is seen sitting bolt upright at the entrance to is burrow, which stays open, unlike the gopher's. It is common in colonies whose members call back and forth.

Grass
Oak
Wd.Pr.
Rocks

POCKET GOPHER (Thomomys bottae and other species). There are several species of pocket gopher within the coastal forest region. Rat - . The two principal ones are the Valley Pocket Gopher (T. bottae) & the Northern Pocket Gopher (T. talpoides), the first being found in California and the

Grass
Brush
Wd.-Pr.

second in Oregon and Washington. The pocket gopher is rarely
seen above ground, and usually shows its presence by a circular
pile of freshly-dug earth pushed out from the underground tunnel.
Moles do this too, but mole burrows usually show narrow ridges
of earth extending for several feet or more. The usually brown-
colored gopher has visible slits which mark the openings into its
external cheek pouches, short ears, and a naked, short, blunt tail.

*Conifer
Brush
Meadow*

APLODONTIA FAM.
(Aplodontidae)

MOUNTAIN BEAV-
ER (Aplodontia rufa).
Cat size. Few animals
are as badly misnamed
as this, which is not a
beaver, and does not
necessarily dwell in
mountains.   It looks
like a large, dark-brown, tail-less gopher, and digs gopher-
like holes below the surface of the ground, but leaves no dirt-
piles at the entrance-ways. Feeds on plants and has colonies.

BEAVER FAMILY (Castoridae)

BEAVER (Castor canadensis). Raccoon size. This rodent is

*Water
Marsh
Hardw.*

almost entirely aquatic
and is a powerful swim-
mer, but slow and
clumsy on land.   Its
large size and broad,
flat tail are distinc-
tive. It cuts down wil-
lows, alders and cot-
tonwoods for food, and
builds dams in the
smaller streams, but
the coastal form rarely
builds the big lodges
that the beavers in the
mountains construct.

Instead it lives in burrows dug in the banks, and in the larger rivers dams are not constructed. The chisel-like teeth of the beaver leave characteristic marks on the felled hardwood trees along the stream banks and show the presence of the animals. Branches and sections of the trunks are often stored underwater.

NATIVE RATS AND MICE FAMILY (Cricetidae)

MUSKRAT (Ondatra zibeth-ica). Rat +. The tail of the muskrat is its most distinctive feature, being long, hairless and strongly flattened from side to side. The tail of the beaver is flattened from top to bottom. The fur is brown. Large, cone-shaped lodges, made of piles of rushes, cat-tails and other aquatic vegeta-tion, are sure signs of musk-rats, but, in some areas, they in-habit holes in banks. Found from Coos Co., Ore., N. into Wash.

Marsh Water

WOOD RAT (Neotoma cinerea and N. fuscipes). In Washington the bushy-tailed woodrat (N. ciner-ea), and in California, the dusky-footed wood rat (N. fuscipes) are the most common species. Both occur in Oregon.

*Conif.
Brush
Oak
Hardw.*

The bushy-tailed wood rat has a squirrel-like tail and whitish hind feet, while the dusky-footed wood rat has a short-haired tail and dark feet. Both wood rats, otherwise known as pack rats, are attractive rodents, with large eyes and ears, and none of the repulsive features of the house rat. Wood rat lodges are commonly cone-shaped piles of short sticks which cover and conceal the nests, feeding chambers, and runways. The rats frequently signal to each other by rattling their tails against dry leaves, but, on the whole, are comparatively solitary animals. They feed mainly on fruits and nuts, and like to collect bright objects.

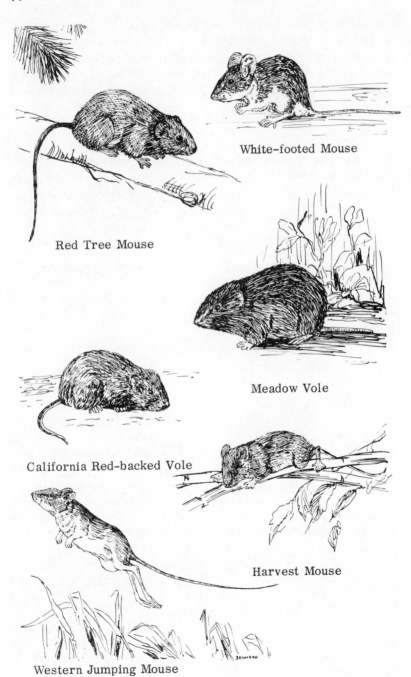

White-footed Mouse

Red Tree Mouse

Meadow Vole

California Red-backed Vole

Harvest Mouse

Western Jumping Mouse

WHITE-FOOTED MOUSE (Peromyscus maniculatus). Mouse size. A large-eared, large-eyed mouse with a long tail, but shorter than head and body combined. The contrasting gray or brownish upperparts, and the clear white underparts are distinctive. It is secretive, coming out mainly at night, and sometimes enters houses. It is more attractive than the brown house mouse. *Most habitats*

RED TREE MOUSE (Phenacomys longicaudus). Mouse size. This is a small, vole-like mouse with short ears, but it differs from the true voles by its longer tail, and by living mainly in the tops of coniferous trees, mainly Douglas and lowland firs, where it builds a large nest, a foot or more in diameter, among the branches. It is more reddish in color than the red-backed vole and has a contrasting blackish tail. It is confined to the coastal strip from north-central Oregon to central California. *Conif.*

MEADOW VOLE (Microtus californicus and related species). Mouse +. There are four or more related species in our region, too closely alike to be separated save by an expert. All are short-tailed, short-eared mice that dwell in tall, dense grasslands where their runway systems form an intricate network both above and below ground. Color brownish-gray. *Wd.Pr. Marsh*

CALIFORNIA RED-BACKED VOLE (Clethrionomys californicus). Mouse size. Dark chestnut-brown in color, with a relatively short tail and ears. The reddish back distinguishes it from the meadow vole, as does its forest floor habitat. *Conif.*

HARVEST MOUSE (Reithrodontomys megalotis). Mouse -. This species resembles a small, white-footed mouse, but the front incisor teeth have distinct, vertical grooves, which the white-footed mouse does not have. It is also more characteristic of grassy areas than is the white-footed mouse. *Wd.-Pr. Strand*

JUMPING MOUSE FAMILY (Zapodidae)

WESTERN JUMPING MOUSE (Zapus princeps). Mouse size. The extremely long tail and large hind feet distinguish this mouse from other species in the region. The ears are smaller than the white-footed mouse. It is essentially a tri-colored mouse with a yellow side stripe separating the brown back from the whitish belly area. It is often seen at night, jumping kangaroo-like in the moonlight across roads or through damp meadows. These mice usually hibernate during the winter. *Meadow Marsh*

Pacific Water Shrew

Trowbridge Shrew

Shrew Mole

Townsend Mole

ORDER OF INSECTIVORES (INSECTIVORA)

THE SHREW FAMILY (Soricidae)

PACIFIC WATER SHREW (Sorex bendirei). Mouse size. All shrews are mouse-like animals, from which they differ external-ly in their smaller size, long-pointed nose, ear-less appearance and small, bead-like eyes. They usually hide themselves secre-tively when they sense the footsteps of a larger animal or a man and come out mainly at night. The water shrew differs from oth-ers in this region in having a row of stiff, bristle-like swimming hairs on the sides of the hind feet. The Northern Water Shrew(S. palustris) occurs in this region only in northern Washington. The northern species prefers cold mountain streams whereas the Pac-ific Water Shrew is more at home in slow streams, but both are expert swimmers and hunters of water insects and larvae.

Water Marsh

TROWBRIDGE SHREW (S. trowbridgei). Mouse -. Other than water shrews there are 5 species of shrews in our region. Of the group the Trowbridge and Vagrant Shrews (S. vagrans) are the more abundant, with the former common among the coniferous forests and the latter preferring marshes, bogs and meadows (as does the Masked Shrew of northern Washington, S. cinereus). Two other species, the Pacific Shrew (S. pacificus) and the small Dusky Shrew (S. obscurus) occur mainly among the conifers. All these animals are brown in color & hunt insects, worms & mice by nose.

Marsh Conif. Hardw. Strand Str.Wd.

THE MOLE FAMILY (Talpidae)

SHREW-MOLE (Neurotrichus gibbsi). Mouse size. This an-imal does not burrow like the true moles, but forms an intricate network of runways deep in the forest litter. It resembles a large shrew, but differs in having a thick and heavy tail with stiff black hairs, unlike the mouse-like tail of a shrew. It differs from the mole in having long, narrow, walking feet.

Conif. Marsh Str.Wd.

TOWNSEND MOLE (Scapanus townsendi). Rat -. Moles are rarely seen, dwelling as they do in either deep or close-to-the-surface burrows. The surface burrows appear as conspicuous ridges of dirt, often many yards long. Moles have broad, strong-clawed, digging feet, little suited for above-ground walks. Their eyes are very tiny and they lack external ears. The fur is vel-vety soft and dark brown or black in color. The California Mole (S. latimanus) and the Pacific Mole (S. orarius) are similar.

Wd.Pr. Hardw. Meadow

ORDER OF BATS (CHIROPTERA)

PLAIN-NOSED BAT FAMILY (Vespertilionidae)

LITTLE BROWN BAT (Myotis lucifugus). Wingspread 8-10".
The little brown bat is a typical example of the Myotis or mouse-
Water eared group of bats, of which 7 species occur in this region, all
Marsh
Wd.Pr. difficult to recognize separately save by an expert. But all are
Brush among the smallest bats in the region and have mouse-like snouts,
tail membranes that are not furred on the top side, and are generally brown in color. Like most bats, they roost by day in caves,
buildings or hollow trees, hanging head downward. In the evening they appear usually later than some of the larger bats, flying
low and erratically, while they chase insects, often over water.

BIG BROWN BAT (Eptesicus fuscus). Wingspread 13". This
Hardw. looks like a large, plain-snouted Myotis, but it lacks fur on the
Conif.
Meadow tail membrane and has relatively small ears. It is usually the
first bat out in the evening, and is a rapid flier, without erratic
or jerky flight. Usually only 1 or 2 are present hunting together.

SILVER-HAIRED BAT (Lasionycteris noctivagans). Wing-
spread 12". This bat is dark brown in color with silvery hair
Hardw.
Conif. in the middle of the back. It is the darkest colored bat in the re-
Meadow gion, and the tail membrane is furred only on the basal half.
The ears are relatively short. It is an early evening flier along
with the Big Brown Bat, but usually flies higher and more errat-
ically, and more often in larger numbers.

Hardw. HOARY BAT (Lasiurus cinereus). Wingspread 16". This is the
Conif. largest bat in the region, and a large, short-eared bat appearing
Meadow late in the evening will usually be this species. White-tipped hairs
over the body give a silvery color. Tail membrane furred all over.

WESTERN BIG-EARED BAT (Corynorhinus rafinesque) and
PALLID BAT (Antrozous pallidus). Wingspreads 12" and 14".
Most
habitats These bats may be distinguished from others by their large ears,
which are usually over 1" in length. The big-eared bat has the
ears joined together over the forehead (as illustrated), while the
pallid bat has its ears separated, and is much paler in color. The
big-eared bat has two large lumps on its nose; the pallid bat does
not have these. The big-eared bat is the more common in this region, while the pallid bat enters the southern part of the region but
does not occur in coastal Washington. Both are cave-dwellers.

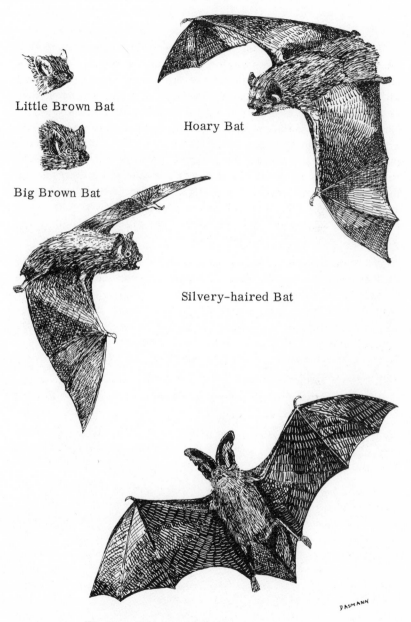

Little Brown Bat

Big Brown Bat

Hoary Bat

Silvery-haired Bat

Western Big-eared Bat

BIRDS

The following pages will help to familiarize the reader with the common birds of this region. It would be difficult in a book of this size to describe even half of the several hundred species which occur in the coastal forest belt. From each major family or order of birds a few of the species most common in, or most characteristic of, the coastal region have been selected. These are the ones that are most likely to be observed. If the characteristics of these species are learned, it will aid in the identification of the other members of the families to which they belong. In many instances only the male bird is illustrated or described. The female will often be seen in the same vicinity as the male, and in most instances will be similar in shape and size, but somewhat less strikingly colored.

Many of the species illustrated are seasonal visitors, abundant at one season, but absent at others, while other birds are permanent residents. As you read the descriptions, you will be helped greatly in identifying the birds by carefully noting the season of the year and the habitats in which each species is commonly found. However, it should be remembered that habitat and seasonal preferences are not always strictly observed and that exceptions sometimes occur to prove the rule.

Size descriptions are given by comparison with the following commonly known birds:

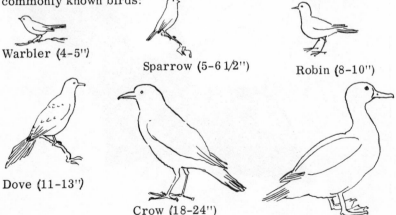

Warbler (4-5")

Sparrow (5-6 1/2")

Robin (8-10")

Dove (11-13")

Crow (18-24")

Mallard (20-28")

ORDER GAVIIFORMES - LOONS

Loons are capable divers, remaining submerged for long periods. They are stream-lined, swimming low in water, duck-size or larger, & with a stout, pointed bill. The similar-looking cormorants are blacker with a longer neck and tail, and more rapid flight.

PACIFIC LOON (Gavia arctica). Mallard size. Common to abundant winter visitor. Straight bill and dark color distinguish it from the Red-throated Loon (G. stellata), which has a slightly upturned bill. Smaller size and thinner bill distinguish it from the Common Loon (G. immer). Feeds on fish, usually in salt water. *Water*

ORDER COLYMBIFORMES - GREBES

Grebes are usually seen swimming or diving in ponds, lagoons and lakes. They are smaller than ducks, with a slender neck, pointed bill and small head. Their feet are lobed rather than webbed as in ducks.

HORNED GREBE (Colymbus auritus). Dove size. The dark back and top of head contrast sharply with the white underparts. The slender neck, small head, and sharp-pointed bill are distinctive. Salt water. *Water*

PIED-BILLED GREBE (Podilymbus podiceps). Dove size. The brownish color and chicken-like, marked bill are distinctive. Resident in ponds. *Water*

ORDER CICONIIFORMES - HERONS

Long-legged, long-necked wading birds, most frequently seen standing along edges of marshes or ponds; commonly, but incorrectly, called "cranes". Fly with head pulled back and neck in S-shape (cranes straight).

Marsh Water Strand Meadow GREAT BLUE HERON (Ardea herodias). About 4' high. Common resident throughout region. The long legs, long neck, blue-gray color, and long, sharp-pointed bill are distinctive.

Marsh Water

AMERICAN EGRET (Casmerodius albus). Slightly smaller than blue heron. A common resident, and a large, all white bird with long neck and long legs. Differs from Snowy Egret (C. thula), which is rare in this region, by larger size and yellow bill. Snowy has black bill. Calif. & S. Ore.

Water Marsh BLACK-CROWNED NIGHT HERON (Nycticorax nycticorax). Crow size plus. A medium-sized heron with black back and whitish under parts. The contrasting black and white on head is distinctive. Commonly seen with neck retracted, giving a large-headed, short-legged, neckless appearance. The immature birds resemble the plain brownish-streaky *Marsh* American Bittern (Botaurus lentiginosus), which is rare in the region, but they differ in gray general color & lack of black wing tips. Resident in Calif., summering to north. Often in salt water.

ORDER ANSERIFORMES - DUCKS, GEESE, AND SWANS

BLACK BRANT (Branta berni-cla). Mallard size and +. Medi-um-sized goose with black head, neck and breast. Small white patch on neck. The related Cana-da Goose (B. canadensis) has a white patch on the cheek & light-colored breast. The brant largely depends on eelgrass for food in winter. Winter visitor, bay shore. *Marsh Water*

MALLARD (Anas platyrhynchos). The males are distinc-tively green-headed with white neck ring. Both sexes have yel-lowish bills and wing patches bordered with white. Female mottled brown. A common resident.

Marsh Water

CANVAS-BACK (Aythya valisineria). Mallard size. Common in winter. Males ap-pear white with reddish head and neck. Has flat profile with forehead sloping strongly back from bill, which dis-tinguishes it from sim-ilar red-head duck. *Marsh Water*

SPOONBILL or SHOVEL-LER (Spatula clypeata). Mal-lard -. A winter visitor. The males appear largely black and white with large pale blue wing patch & large, flattened spoon-like bill. Reddish belly & sides.

Marsh Water

Marsh Water

LESSER SCAUP (Aythya affinis). Mallard -. The high forehead is distinctive; appears blackish on head, breast and tail; white to gray on wings and underbelly. Females less brightly marked and with white area at base of bill. Abundant winter visitor on bays, estuaries and adjoining marshes.

Marsh Water

AMERICAN MERGANSER (Mergus merganser). Mallard size. Mergansers differ from other ducks in having slender bills with toothed edges. They feed on fish. Males of this species are largely white, with black head and back, and an orange bill; females have crested red-brown heads & red bills. Resident.

Marsh Water

RUDDY DUCK (Oxyura jamaicensis). Mallard -. One of the smaller ducks. White cheeks, dark crown, and up-turned, spiny tail feathers are distinctive. It dives almost as readily and frequently as do loons and grebes. Resident, on ponds, lakes and coastal bays.

Marsh Water

BALDPATE (Mareca americana). Not illustrated. The male is quite distinctive, with a gray head topped by a white crown, a glossy-green patch on the side of the head being sometimes visible in good light; the body appears mainly brownish. The female is more reddish-brown, with gray neck and head. In flight both sexes can be recognized by the large white patch that shows clearly on the front edge of the wing. Mainly winter visitor.

Marsh Water

PINTAIL (Dafila acuta). A long slender neck, a long pointed tail, a white line on neck, and white belly mark the male. Res.

ORDER FALCONIFORMES
HAWK-LIKE BIRDS

TURKEY VUL-
TURE (Cathartes
aura). Crow size
and +. 6' wing
spread. Common
in summer; most
often seen soaring
with wings lifted
at an angle above
horizontal, while
eagles hold wings

Strand
Wd.Pr.
Rocks
Brush

horizontal from body. The black color combined with the naked,
reddish head and narrow tail are dis-
tinctive. Feeds on carrion.

SHARP-SHINNED
HAWK (Accipiter stri-
atus). Dove size. The
Accipiter hawks have
short, rounded wings,
and a long tail. This
is the smallest & com-
monest of 3 species in
the region, and has a

Conif.
Hardw.
Wd.Pr.
Brush

square-ended tail. The Cooper Hawk (A. cooperi) is larger than
the sharp-shin, but is best distinguished by the rounded end to the
tail. The Goshawk (Astur atricapillus) is nearly twice the size
of the Cooper & lighter gray. All eat birds.

RED-TAILED HAWK (Buteo jamaicensis). Crow size
and +. The Bu-
teo hawks have
short tails, usu-
ally spread fan-
like in flight, and
long, blunt-ended
wings. Red-top-
ped tail distinc-
tive. Resident.

Strand
Wd.-Pr.
Meadow

MARSH HAWK (<u>Circus</u> <u>cyaneus</u>).

Marsh Meadow Wd.Pr.

Crow size. Long, rounded wings combined with a long, square-ended tail and <u>a conspicuous white rump patch</u> serve to distinguish this species. The old males are pale grey, almost white; the females brownish.

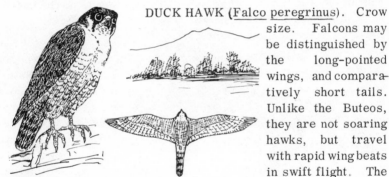

Low flight. Common in summer in north; resident in California.

DUCK HAWK (<u>Falco</u> <u>peregrinus</u>).

Rocks Marsh Water Strand

Crow size. Falcons may be distinguished by the long-pointed wings, and comparatively short tails. Unlike the Buteos, they are not soaring hawks, but travel with rapid wing beats in swift flight. The duck hawk, or Peregrine Falcon, is the largest falcon in the region, and is distinguished by its dark, nearly black color above, and the <u>dark, moustache pattern on its face</u>. Resident; hunts mainly the large birds, such as ducks; very fast flight.

SPARROW HAWK (<u>Falco</u> <u>sparverius</u>). Robin +.

Strand Meadow Wd.Pr.

The sparrow hawk or kestrel is the smallest American falcon. It is distinguished by the <u>bright, reddish-brown color of the tail</u>, and by its tendency to hover with rapid wing beats, while hunting mice & grasshoppers. Call a rapid "killy-killy-killy!" Often lights on telephone wires.

E.D.

ORDER GALLIFORMES - CHICKEN-LIKE BIRDS

BLUE GROUSE (Dendragapus obscurus). Crow size and +. A large, dark, blue-gray, chicken-like bird, with a light band at the tip of its tail. Often seen on the ground, or high in coniferous trees. The dull, muffled, hooting call of the blue grouse is often heard in the spring. Resident.

Conif. Hardw. Wd.Pr.

RUFFED GROUSE (Bonasa umbellus). Crow size & +. The uncommon ruffed grouse is distinguished from the above species by its red-brown color, and the black band near the tip of the tail. The drumming call of the male is distinctive. From Mendocino Co. north into British Columbia; most common in burnt-over land.

Brush Hardw. Conif.

MOUNTAIN QUAIL (Oreortyx picta). Dove size & +. Resident and locally common. Distinguished by bright chestnut and blue-gray color, white barring on flanks, and long, straight plume on the head. Usually found in small flocks, called coveys, and characteristic of more mountainous areas. Gives a loud, mellow "wook" or "t-wook" cry.

Brush Wd.Pr. Hardw. Conif.

*Brush
Wd.-Pr.*

CALIFORNIA QUAIL (Lo-phortyx californica). Dove size. Smaller than mountain quail, less brightly colored, & with short, forward-curving head plume. Common resident from southern Oregon south; found in large or small coveys. Gives "quer-ca-go!" cry; female calls young with "pt-pt-pt!"

ORDER GRUIFORMES - CRANES, RAILS, ETC.

*Water
Marsh
Meadow*

AMERICAN COOT (Ful-ica americana). Mallard -. This duck-like bird with a whitish, chicken-like bill is slate gray to black color all over.   It is commonly seen swimming with head moving back and forth; a slow, clumsy flier. Per-manent resident, abundant, often in large flocks.

ORDER CHARADRIIFORMES - SHORE BIRDS & GULLS

The shore birds include the sandpipers, plovers, gulls, terns and their relatives. They are characteristically birds of the sandy beaches, mud flats, or rocky ocean shores. At least 45 species occur within this region, including many forms difficult for the beginner to distinguish.   Of these we have illustrated only six representative species.

*Strand
Marsh
Water
Meadow*

KILLDEER (Charadrius vo-ciferus). Robin size.   This plover is a common resident of the region. The even, dark-brown colored upper-parts, and the two dark breast bands are distinctive, as is the loud, sharp "ki-deer" or "ki-dee" call. Often nests on river bars.

WILSON'S SNIPE (Ca-pella gallinago). Robin +. The long slender bill, color pattern and size distinguish the "Jack snipe". When alarmed, it flies off in a sharp and zig-zagging flight. It probes in mud for insect larvae, etc. Resident.

Water Marsh Meadow

LONG-BILLED CURLEW (Num-enius american-us). Crow +. Bill is 5-7" long. Curlews are known by their long, slender, strongly down-curved bills. Two species oc-cur in this re-gion, of which this is the larg-er. Migrant and winter visitor.

Water Marsh Strand

WILLET (Catoptrophorus semipalmatus). Dove +. The relatively large size, long straight bill, and even brownish-gray color are distinctive when the bird is on the ground. In flight, a striking black and white wing pattern is displayed. The legs are blue-gray in color. Common migrant and wintering bird on the beaches. Gives a loud "Kay-ti!" call in winter.

Water Strand

Marsh
Water
Strand

DOWITCHER (Limno-dromus griseus).   Dove size.   A squat, long-billed sandpiper, with legs much shorter than the willet. Much heavier in build than the Wilson's snipe.   The lower back and tail are white. Com-mon migrant in fall and winter.   Call, a thin "kleeek", often repeated.

Water
Strand
Bldg.

WESTERN   GULL (Larus occidentalis). Size of crow and +.   Of the 10 species of gulls which occur in fair numbers within this region, the Western may be recog-nized as one of the larger, with a dark, almost black back, a white head, and white underparts. Other gulls of similar size are light gray above.   This is the only common resident gull over all the region, found mainly on or near salt water.

Water
Strand
Bldg.

GLAUCOUS-WINGED GULL (Larus glaucescens).   Crow +. Equal in size to the western Gull, but lacking in black coloring.

Pale gray above, white below, with wing-tips marked with gray rather than black.   Feet pink. All gulls are confusing when immature, being variously marked with gray or brown. The white underparts and head may not appear until the third year. Breeds along Wash. coast; winters all region.

ORDER COLUMBIFORMES - PIGEON-LIKE BIRDS

BAND-TAILED PIGEON
(Columba fasciata). Dove +.
Similar in size to the do-
mestic pigeon. Tail fan-like,
and with broad pale band
near tip. White crescent on
nape of neck. General col-
or is blue-gray. Travels in
flocks of varying size, and
gives owl-like "whoo-ooo-
ooo" cry. Summer resident.

Oak
Hardw.
Conif.

MOURNING DOVE (Zenai-
dura macroura). Smaller
than the pigeon, more slim in
build, and with a long, pointed tail. Often solitary, or in
pairs, but gathering in larger
aggregations during migra-
tion. The tail feathers flash
white in flight and the wings
whistle. A mournful "coo-
coo-coo" cry is heard.

Wd.Pr.
Hardw.
Str.Wd.

ORDER STRIGIFORMES - OWLS

SCREECH OWL (Otus asio).
Recognized as a small, fat-look-
ing owl, with conspicuous ear-
tufts. Other small owls in this
region lack ear tufts. Usual col-
or is gray, but one race is brown.
Nests in tree cavities. Resident
and fairly common in broken coniferous forest, or stands of
broadleaf trees. Voice not really
a screech, but a series of soft,
tremulous whistles, often running together at the end.

Conif.
Brush
Hardw.
Wd.Pr.

Wd.Pr.
Hardw.
Brush
Conif.

GREAT HORNED OWL (Bubo virginianus). Crow +. A large owl with conspicuous horn-like ear tufts. In flight all owls differ from hawks by large-headed neckless appearance, and soft, noiseless wing beats. The horned owl is larger and more heavily bodied than the redtailed hawk, and is likely to be seen only at dusk or early morning. A fairly common resident; an active killer of harmful rodents and rabbits. Large, grayish pellets at the base of a tree, each made up of hair and bones, tell of the owl's presence and the food it eats. A deep "hoo hoo!"

Conif.
Hardw.

SPOTTED OWL (Strix occidentalis). Crow size. A large, earless owl with heavily spotted and barred underparts. Instead of the deep hoot of the horned owl, this bird has a high-pitched "hoo-whoowhoo-ooo" that may sound like a dog barking. Resident and fairly common in the region.

ORDER CAPRIMULGIFORMES
NIGHTHAWKS & RELATIVES

Wd.Pr.

PACIFIC NIGHTHAWK (Chordeiles minor). Robin size. Most often seen flying at dusk over open areas. Flight erratic, as it darts after insects. Wings narrow, pointed, marked with conspicuous white spot near the tip. Throat white, underparts and tail barred. Nighthawks have weak legs, and rest on ground or broad limbs when in flight. A loud "peent" call advertises its presence. Roosts in open coniferous forest; flies over open country catching insects, generally in the evening. A summer resident.

ORDER APODIFORMES - SWIFTS & HUMMINGBIRDS

VAUX SWIFT (Chaetura vauxi). Warbler size. This is a small, cigar-shaped bird with long, slightly curved, stiff wings. Wings more stiff and slender than those of swallows. General color dark; tail inconspicuous. A summer visitor, flying high in the air to catch insects, roosting at night in high trees.

Conif.
Hardw.
Meadow

RUFOUS HUMMINGBIRD (Selasphorus rufus). Warbler -. Hummingbirds are the smallest of birds, marked with irridescent colors, and characterized by a wing beat so rapid that the wings are almost invisible. Often seen hovering in front of flowers and probing them with a long, tube-like bill. 2 species occur commonly in this region. The rufous is the only one with a bright, red-brown back; the Allen Hummingbird (S. alleni) has a green back and is found from Humboldt Co. S. Males of both have bright, reddish-orange throats. The Anna (Calypte anna) of S.F. Bay has red throat.

Brush
Cultiv.

Conif.
Brush
Oak

ORDER CORACIIFORMES
KINGFISHERS, ETC.

BELTED KINGFISHER (Megaceryle alcyon). Robin +. The blue-gray color, large head, stout bill, and uneven crest serve to distinguish this bird. Gives a rattling call in flight; feeds on small fish.

Water
Str.Wd.
Cultiv.

ORDER PICIFORMES - WOODPECKERS

Hardw.
Oak
Str.Wd.
Conif.

DOWNY WOODPECKER (Dendrocopus pubescens). Sparrow +. The smallest woodpecker in the region. The white back, spot of red on black and white head, and small, weak bill distinguish it from others. The similar Hairy Woodpecker (D. villosus) is larger and has a stronger bill. The downy often clings to branches and works around them, often upside down, seeking insects.

Hardw.
Oak
Conif.

YELLOW-BEL-LIED SAPSUCKER (Sphyrapicus varius). Robin -. Red crown and throat, black and white wings and back, and yellow belly are distinctive. Females have white throat; however the broad expanse of red on the head is characteristic. Resident. Has a nasal, churring cry.

Wd.-Pr.
Hardw.
Conif.

RED-SHAFTED FLICKER (Colaptes cafer). Dove size. The only woodpecker regularly seen on the ground where it feeds on ants, etc. Brown back, red under wings and tail, white rump, and black crescent below throat are distinctive.

Conif.

PILEATED WOODPECKER (Dryocopus pileatus). Crow size. This is the largest and most striking woodpecker in the region, with large head and conspicuous red crest; a black and white pattern shows in flight. Resident.

ORDER PASSERIFORMES - PERCHING BIRDS

Flycatcher Family, Tyrannidae

OLIVE-SIDED FLYCATCH-
ER (Nuttallornis borealis).
Sparrow +. Flycatchers are
distinguished by their habit of
seeking high or exposed perches
from which they dart out into the
air, capture insects on the wing,
and return to their perch. The
olive-sided flycatcher is the
largest of the family in the region, and is more commonly
heard than seen, with a distinctive three-note call, with middle note highest. Distinguished
also by large head and bill, white

Conif.
Hardw.
Meadow

throat and dark side patches separated by white belly. Summer.

WESTERN WOOD PEWEE (Contopus richardsonii). Sparrow size. A
gray-brown flycatcher. Two white
wing bars, and absence of white eye-
ring separate it from olive-sided flycatcher, which lacks wing bars, and
the smaller western flycatcher, which
has a conspicuous eye ring. Cry, a
nasal "pee-eee". Summer visitor to
forest edges and open forest.

Conif.
Hardw.
Oak

WESTERN FLYCATCHER (Em-
pidonax difficilis). Sparrow -
This is a small flycatcher with a
conspicuous eye ring, two white
wing bars, and yellowish underparts. Call, a sharp "see-seet" or
"ps-seet", rising on second note.
Common in summer, particularly
near streams or forest edges.

Conif.
Hardw.
Str.Wd.

Swallow Family (Hirundinidae)

Wd.Pr.
Meadow
Rocks
Bldg.
Strand

E.D.

VIOLET GREEN SWAL-
LOW (Tachycineta thalas-
sina).   Sparrow size.
Swallows are most often
seen in flight.   The long,
slender wings, and gliding,
easy flight are distinctive.
The green and purple up-
perparts,   clear white underparts, white patches to
side and above the tail, and
the white extending partly around the eye are distinctive in this
species.   Summer resident,
nesting on cliffs, buildings;
hunting insects in open. Simi-

Meadow
Conif.
Hardw.

lar Tree Swallow (Iridoprocne
bicolor) lacks white patches.

Conif.
Meadow
Hardw.

PURPLE MARTIN (Prog-
ne subis). Robin -. This is
the largest of the swallows, and
only one dark blue-black all
over in male; brown in female,
with pale collar and whitish
belly.   Summer   resident;
nests in tree holes in open forest.

Crow & Jay Fam. (Corvidae)

Conif.

CANADA JAY (Perisoreus
canadensis). Robin +. A large
gray bird with a black bill and
white head; underparts whitish.
Has many harsh & soft cries,
including a soft whistled "whee-
ahh".   Young are dark gray. A
resident from Mendocino Co. N.

Wd.-Pr.
Oak

E.D.

SCRUB JAY (Aphelocoma
californica). Robin +.   Blue
head, wings & tail; no crest.

STELLER JAY (Cyanocittá stelleri). Robin +. This is a large bird, black in front, blue toward rear, with prominent black crest. Common resident. Call a hoarse "tchay-tchay-tchay!, also many other calls, in-cluding copy of red-tail hawk scream. Resident.

Conif.
Oak
Hardw.

RAVEN (Corvus corax). Crow +. This huge, all black bird, is nearly twice as big as a crow. Unlike the Crow (C. brachyrhynchos), it soars with wings held horizontal from body (not angled). Croaking call differ-ent from "caw" of crow. Resident.

Wd.-Pr.
Strand
Rocks

Str.Wd.
Strand
Wd.Pr.
Hardw.

Titmouse Family (Paridae)

CHESTNUT-BACKED CHICKA-DEE (Parus rufescens). Warbler size. The black and white face mask and small size are distinc-tive; also the reddish-brown back.

Hardw.
Conif.

BUSH-TIT (Psaltriparus mini-mus). Warbler -. A small, gray bird, traveling in flocks, except at nesting time. Lacks distinguishing marks other than small size, uni-form color, rather fat body, and long tail. Common resident.

Hardw.
Oak
Brush

Nuthatch Family (Sittidae)

Conif.

RED-BREASTED NUTHATCH (Sitta canadensis). Warbler size. Nuthatches are small, short-tailed birds that are often seen walking up and down vertical tree trunks, or clinging upside down from large branches. This species is distin-

guished by reddish underparts and a black line through the eye.

Strand
Conif.

It is blue-gray above, light-colored below, black capped. The Pygmy Nuthatch (S. pygmaea), has gray-brown cap. Mendocino S.

Creeper Family (Certhiidae)

Conif.

BROWN CREEPER (Certhia familiaris). Warbler size. Has striped brown back, a stiff, short tail, and a curved bill. It creeps spirally up vertical tree trunks then flies to bottom of next tree. It never climbs down a trunk as does a nuthatch. Has a whispering "seeee" note. Resident.

Wren-tit Family (Chamaeidae)

Brush

WREN-TIT (Chamaea fasciata). Sparrow size. A brownish bird, dark above, with a short, slightly curved bill, and a long tail, usually held upright. N. to Ore.

Wren Family (Troglodytidae)

Brush
Hardw.
Oak.

BEWICK WREN (Thryomanes bewicki). Warbler size. A wren with a white eye line, and white edge to end of tail; the tail rather long. Resident in most of region.

WINTER WREN (<u>Troglodytes</u> troglodytes). Warbler size. Wrens are characterized by narrow, stiff tails, which are
usually held upright, and a relatively long,
down-curved bill. This species is dark
brown, with a <u>heavily-barred flank & belly.</u>
Has a light, but not white, line above eye; a
very short tail. Resident. The House Wren
(T. aedon) has longer tail, unbarred flanks, no eye marks. Res.

*Brush
Hardw.
Conif.*

*Oak
Str.Wd.
Brush
Hardw.*

Dipper Fam. (Cinclidae)

DIPPER (<u>Cinclus</u> mex-
<u>icanus</u>). Robin -. A
short-tailed, heavy-bod-
ied, slate-gray bird.
<u>Dips up and down</u> while
standing on rocks by swift
water; runs and flies un-
der water hunting water
insects. Resident in mts.
Nests of moss often hid-
den under waterfalls.

*Water
Str.Wd.
Rocks*

Thrush Fam. (Turdidae)

ROBIN (<u>Turdus migra-
torius</u>). A familiar bird of
the city lawn and shade tree.
Breast is orange-red, body
is brown. Males are black-
headed; females more
brownish or paler on the
head.

MEXICAN BLUEBIRD (<u>Sialia mexi-
icana</u>). Sparrow +. Mostly blue, with
reddish breast and back; sometimes
back also is blue; the female is dull
brownish above and is otherwise paler
than the male. Voice, a sharp "tew!"
or "mew!"; also has loud chatter.

*Cultiv.
Meadow
Wd.-Pr.
Conif.*

*Wd.Pr.
Hardw.
Meadow
Conif.*

Conif.

VARIED   THRUSH
(Ixoreus naevius). Rob-
in size. This is a robin-
like bird with an orange
stripe above the eye and
a black band across the
breast; the female has a
gray breast band, while
young birds have a brok-
en breast band. Has a
long, whistled note that
quavers eerily. Resid.

Conif.

HERMIT THRUSH (Hylocich-
la guttata). Robin -.   Grayish-
brown above, with spotted breast
and reddish tail. Runs on ground,
or often stands still while watch-
ing for insect prey, slowly bob-
bing tail. Has a lovely, clear and
flute-like song. Resident. The
Russet-backed Thrush (H. ustu-

Str.Wd.
Hardw.

lata) looks the same but has no
reddish tail; a summer visitor, with lovely song that spirals down.

Conif
Hardw.
Oak
Brush

Kinglet Family (Sylvidae)

GOLDEN-CROWNED   KING-
LET (Regulus satrapa). Warbler
-. This is a very small gray bird
with a short tail, and black and
gold stripes on crown. Resident.

Conif.
Hardw.
Oak
Brush

RUBY-CROWNED KINGLET (Reg-
ulus calendula). Warbler -. A short-
tailed grey bird with white eye ring,
and two white wing bars; the bright
red crown patch is usually unseen.

Waxwing Fam., Bombycillidae

CEDAR WAXWING (Bombycilla ce-
drorum). Sparrow +. Waxwings are
crested birds with black masks and
yellow bands at the ends of their tails.
The cedar waxwing is brownish in col-
or; the rarer Bohemian Waxwing (B.
garrula) is larger, grayer, and has
white patches on the wings, visible as
the wings are folded. Resident, feeding on berries; breeds by streams, &
winters in brushy areas or woodland.

Brush Hardw.
Str.Wd.
Wd.Pr.

Vireo Family (Vireonidae)

HUTTON VIREO (Vireo hut-
toni). Warbler size. Somewhat
like a kinglet, but larger, more
olive in color, and slower and
more deliberate in movements.
Two white wing bars, light spot
in front of eye, darker area on
top of eye, and incomplete white eye ring are distinctive. Resident.

Oak

Warbler Family (Parulidae)

YELLOW THROAT (Geothly-
pas trichas). Males are yellow
birds, brownish backed, with
black eye masks. Females and
young without mask. Song, a
very sharp "whitchity-t-whichi-
ty-t-whichity- etc." Resident.

Marsh Hardw.
Str.Wd.

PILEOLATED WARBLER (Wil-
sonia pusilla). Male is yellow, with
black cap on top of its head; black
eyes contrasting with bright yellow
face. Song a wheezy "chee-chee-
che-chit, etc.", becoming faster
and louder. Summer resident.

Brush Str.Wd.
Hardw.

ORANGE-CROWNED WARBLER
(Vermivora celata). A dull-colored

Brush Oak
Str.Wd.
Hardw.

warbler, with weak, trilling song.

Conif.
Hardw.
Oak

TOWNSEND WARBLER (Den-
droica townsendi). A black, white
& yellow warbler, with distinctive
black & gold pattern on head and
bright yellow breast. Winters in
Calif., summer to the north. The

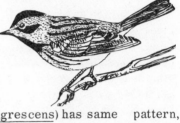

Oak
Hardw.
Conif.

Black-throated Gray Warbler (D. nigrescens) has same   pattern,
but colors all black, white & gray.

Conif.
Hardw.
Op.Brush
Meadow

MYRTLE WARBLER (Den-
droica coronata). This and the
similar Audubon Warbler(D. au-
duboni) are black and white warblers with yellow crown & rump
patch. The myrtle has a white
throat; Audubon a yellow throat.
Myrtle is often seen on ground.
Summer in forest; win. in open.

Blackbird Fam. (Icteridae)

Marsh

RED-WINGED   BLACKBIRD
(Agelaius phoeniceus).   Robin
size.   The male is a shiny black
bird with a red shoulder patch,
the female is streaked and brownish. Call, a loud "keck!" or "tear".
Song, a gurgle-like "kunk-el-o-
ree" or "ok-la-er-ree!"   Often
builds nests among cattails in
fresh-water marshes.

E.D.

Marsh
Grass
Cultiv.

BREWER BLACKBIRD (Eu-
phagus cyanocephalus). Robin
size. The male is all black
with a purple or greenish iri-
descent sheen, and white eyes.
The female is dull brownish.
Resident in marshes in summer; in grassland in winter.
Call, a harsh "keck!"

E.D.

Tanager Family (Thraupidae)

WESTERN TANAGER (Piranga ludoviciana). Sparrow +. The black-winged, yellow-bodied, red-headed male is unmistakable. The female is yellow below and greenish above. Both have heavy, almost finch-like bills. Summer.

Conif. Hardw.

Finch Family (Fringillidae)

All finches have short, thick bills, adapted for seed-cracking.

BLACK-HEADED GROSBEAK (Pheucticus melanocephalus). Sparrow +. A large finch with thick, heavy bill and reddish-brown breast. The male has black and white wings and black head. Summer resident.

Conif. Hardw.

EVENING GROSBEAK (Hesperiphona vespertina). Robin size. A large-billed bird with black and white wings. Body uniformly yellowish in color. Song is a short, wobbling warble. Resident, breeding in dense forest, wintering where berry trees and bushes occur. A camp scavenger in summertime.

Conif. Hardw. Brush

PURPLE FINCH (Carpodacus purpureus). Sparrow size. Male has a reddish head, breast and rump on an overall brownish body. The male House Finch or Linnet (C. mexicanus) is similar, but has dark streaks on the belly and does not have the deeply-notched tail of the purple finch. The females of both species are plain-looking, streaked brown birds. The house finch is found in more open country in summer and its song is much looser and less compact. Residents.

Conif. Brush Oak

Brush Wd.-Pr. Meadow Cultiv. Bldg.

Conif.
Hardw.
Wd.Pr.

PINE SISKIN (Spinus spinus). Warbler +. A goldfinch-like bird, but with yellow only in wing and tail. The general appearance is of a small, brownish, streaked finch. Three calls, a loud, clear "chee-eep", a soft "ti-ti-teet", and a very buzzing "shzzzreeee!" Resident, nesting in conifers; winter feeder in more open country, such as woods' edges.

Hardw.
Str.Wd.
Brush
Wd.Pr.

AMERICAN GOLDFINCH (Spinus tristis). Warbler +. Male in summer recognizable as a yellow bird with a black forehead, wings, and tail. Female, and male in winter, are greenish-yellow with black wings. It is resident, usually nesting along the streams; forages in brush, etc.

Conif.

RED CROSSBILL (Loxia curvirostra). Sparrow size. The male has a red head and body, with darker brownish wings and tail. Bill slender for a finch, with mandibles crossed near tip. Summers to north and in higher mountains, winters southward. The call is a harsh "pip-ip-ip". Feeds on seeds and nuts.

Brush
Oak
Hardw.
Conif.

SPOTTED TOWHEE (Pipilo maculatus). Robin -. Male with black head; neck and back marked with white spots. Sides are red-brown, underparts white. Females similar in pattern, but with brown replacing black. Gives a long, buzzing "chzzzeee!" and also a meow-like call. Common resident. The Brown Towhee (P. fuscus)

Brush
Oak
Hardw.

is found as far north as S.W. Oregon, and is all brown save for rufous under tail.

OREGON JUNCO (Junco oregan-us). Sparrow size. Juncos are characterized by a blackish head, neck and upper breast and by the white-edged tail. This species has a reddish-brown back and sides, and a white belly. Has a soft clicking call; also soft twittering. Summers in conifers; winters in open.

Conif.
Brush
Hardw.
Wd.Pr.

CHIPPING SPARROW (Spizella passerina). A small sparrow with an unmarked grayish breast, a red-brown crown, a white line over the eye, and a black line through the eye. The song is a wheezy, rattling trill, all at one level; call, a dry "cht". Summer.

Hardw.
Oak
Conif.
Wd.Pr.

WHITE-CROWNED SPARROW (Zonotrichia leucophrys). This is a medium-sized sparrow with an unmarked grey breast and alternating black and white stripes on the crown. Song starts with 3 or 4 bright, plaintive whistles, followed by a sneeze-like trill. The first note is often long and quite high, and the rest go down in scale. However, the song varies greatly in the south.

Wd.Pr.
Brush
Strand
Grass

GOLDEN-CROWNED SPAR-ROW (Zonotrichia atracapilla). It is similar to white-crowned sparrow, but with broad yellow stripe on center of head and without a light line over the eye. The three-note song, sounding like the tune of "Three Blind Mice", is very distinctive. This and the above species of sparrow go in small to large flocks. Winter visitor.

Brush
Hardw.
Strand

Conif.
Brush
Hardwd

FOX SPARROW (Passerella iliaca). A large sparrow with a heavily-spotted breast, and an evenly-brownish colored back and wings. Usually seen scratching vigorously in leaf litter. It is chiefly a winter visitor in the more open forests or brush.

Brush
Wd.-Pr.
Hardw.
Marsh
Strand

SONG SPARROW (Melospiza melodia). This is a sparrow with a heavily-streaked breast with a dark central spot, and with dark chin streaks on each side of bill. The back and wings are brownish. The beautiful song usually starts with "sweet-sweet-sweet cheer" and continues with a variety of both musical and buzzy notes. It is an abundant resident in open shrubby areas near marshes and streams, and is common also

Wd.-Pr.

in urban and agricultural areas. The Lincoln Sparrow (M. lincolni), which lives mainly along streams and in marshy spots, has finer streakings on the breast, not so obvious a central spot, and buff-colored band on the upper breast. Song has swift, upward pitch, and gurgles sweetly. Resident in north; winter in south.

Wd.Pr.

SAVANNAH SPARROW (Passerculus sandwichensis). Not illustrated. Similar to song sparrow, with yellow over eye, and a notched tail, but with a shorter tail, a whitish stripe through the crown, and light pinkish legs. Song, soft and whispering "tseœ tseeeeee". Resident in open country.

REPTILES AND AMPHIBIANS

Only a few representatives of the reptiles and amphibians are illustrated in this book, because of space limitations. However, we have inserted two charts, one for reptiles and the other for amphibians, which show most of the species found in this region in relation to the habitats in which they live. Since the living habits of these animals in this region are often very specialized, this information will help you identify almost any species of reptile or amphibian you encounter, if you carefully note the type of habitat in which it is found. The interested reader will find further information about these two groups in the references listed on page 109. Compared to the numbers of species of birds and of mammals, there are few kinds of reptiles and amphibians in this region. But they are easier to catch and keep in captivity.

Reptiles may be distinguished from other vertebrates by their dry, scale-covered skin, which lacks hair or feathers. Their many other distinctive characteristics are not always apparent to casual observation. If it is remembered that snakes, lizards, crocodiles, and turtles are reptiles, there will be little difficulty in distinguishing this group.

Amphibians, like reptiles, represent a distinct class of vertebrate animals, equal in rank to the mammals or birds. As the name indicates, they are creatures which have not completely adapted themselves to life on dry land, but require water or moist places in which to lay their eggs, or in which to spend the early "tadpole" stage of their existence. A characteristic of amphibians is the ability to breath through the skin. In some species this is a secondary method of respiration, but in other groups lungs may be entirely absent.

In the pages that follow we have placed first the chart of the reptiles, then descriptions of outstanding reptile species, next the chart of the amphibians, followed by descriptions of the most common amphibians. As you find reptiles or amphibians in the wild, note their habitats and appearances and compare with the information in the charts plus the pictures of representative species in order to identify them. Most of these creatures may be easily raised in captivity if sufficient care is given to them to make their quarters homelike and their food representative.

CHART OF REPTILE HABITATS AND CHARACTERISTICS

Name	Appearance	Habitats	Range
PACIFIC POND TURTLE (Clemmys marmorata). Illus. on page 102	5-7" long; dark brown, olive to blackish in color above, yellowish below.	Prefers quiet water of small lakes, ponds, slow streams, rivers and marshes.	From N. Ore. S., but avoids Calif. coast from Del Norte to Mendocino.
PAINTED TURTLE (Chrysemys picta).	4-9" long; general brown color, lined and bordered with yellow; red marks on head.	Quiet, shallow waters of rivers, marshes, ponds, ditches, small lakes, etc.	Along Columbia and lower Willamette Rivers
COAST HORNED LIZARD (Phrynosoma coronatum).	3-4" less tail; 2 rows of spines at sides of body; 2 long horns at back of head.	Grass, brush, and edges of hardwood and coniferous forests.	From Sonoma Co. to Monterey Co., Calif. & S.
WESTERN FENCE LIZARD (Sceloporus occidentalis).	2 to 3 1/2" body and head length; tail a little long; scales above eyes separated from central head scales by row of small scales.	Likes wooded, rocky areas, especially in canyons or along a stream; also talus slopes, old buildings, fences, wood piles, burrows of rodents, brush.	From Seattle south, avoiding the coast of Washington and northwest Oregon.
WESTERN SKINK (Eumeces skiltonianus).	2 1/2 to 3 1/2" body and head length; tail around 1 1/2 X as long. Brownish color with light and dark stripes down sides; the scales very smooth. Young have blue tails.	Found under rocks, logs, surface litter, inside rotten wood; in hardwoods, woodland-prairie, brush, conifers. Prefers some shrubs nearby.	From central and S.W. Ore. south to Monterey Co., Calif. & south.
NORTHERN ALLIGATOR LIZARD (Gerrhonotus coeruleus). See illus. on p. 102	3-5" body and head length; described and pictured on page 102. Has dark lines on edges of belly scales.	In or near coniferous forests; also in nearby meadows, hardwoods and brush; often under damp debris or leaves.	In all of region except major part of Willamette Valley in Oregon.
FOOTHILL ALLIGATOR LIZARD (Gerrhonotus multicarinatus).	4 to 6 1/2" body and head length. Like above lizard, but lighter colored, and with dark lines in middle of whitish belly scales.	Usually found in oak wood or brush, but also in grass; usually under debris, logs, leaves, etc.	From interior valleys of Ore. south and then along California coast.
RUBBER SNAKE (Charina bottae). See illustration on page 102.	About 1-2' long; pale brown to dark brown color; tail so blunt it looks like head; no apparent neck.	Usually found in damp localities, in or near coniferous forests; often found under rocks or debris or burrowing.	Found in most of region except far W. Washington and N.W. Oregon.
WESTERN RATTLESNAKE (Crotalus viridis)	1 1/2 to 5' long; easily told by rattles, long fangs, & arrowhead-shaped head.	Brush, woodland-prairie; especially near or in rocks.	From Douglas Co., Ore. and Humboldt Co. S.

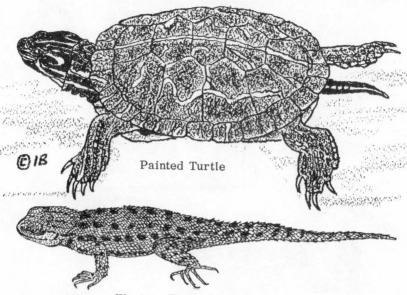

Painted Turtle

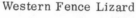

Western Fence Lizard

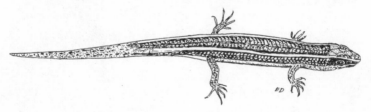

Western Skink

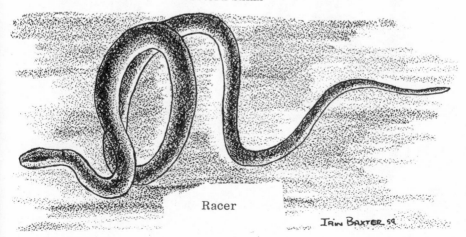

Racer

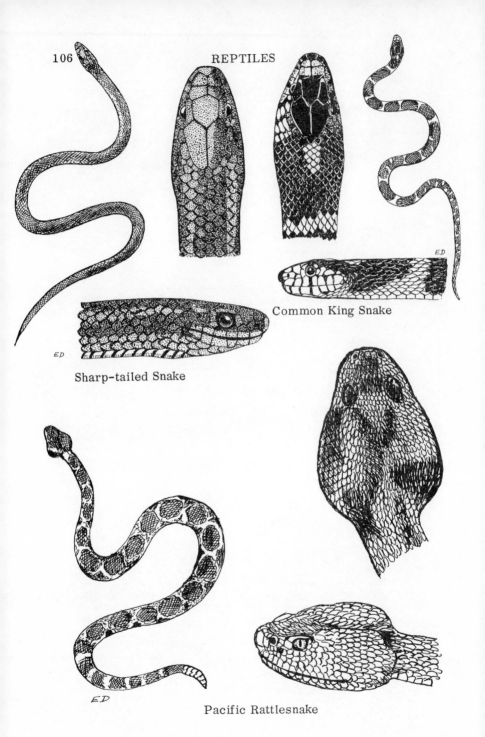

Common King Snake

Sharp-tailed Snake

Pacific Rattlesnake

CHART OF REPTILE HABITATS (continued)

Name	Appearance	Habitats	Range
RACER (Coluber constrictor).	2 to 4 1/2" long; uniformly brownish or dark color above, white below. Very active and fast. Young are saddle-marked.	Woodland-prairie, meadows, edges of forests, in thin brush; sometimes climbs in bushes.	Tacoma, Wash. area; interior valleys of Ore.; S. along coast from Curry Co.
SHARP-TAILED SNAKE (Contia tenuis).	10-16" long; rather stout body, with sharp scale at end of tail; belly appears with black and white cross bars; brown to yellowish-gray on upper. body.	Usually in moist places, under bark, boards, etc., near a stream; also in oaks, brush, hardwood forest.	From Medford, Ore., and Del Norte Co.,California, south.
WESTERN RING-NECKED SNAKE (Diadophis amabilis).	10-22" long; uniformly blackish-gray, or blue-gray, etc. above; with distinct neck ring of yellow, red or orange, and belly of same light color.	Found in coniferous forest, oaks, woodland prairie, hardwood forest; usually under debris, boards, leaves, etc.; rare in marsh.	Found from Portland, Ore. south, avoiding northwest Ore. coast district.
CALIFORNIA MOUNTAIN KING SNAKE (Lampropeltis zonata).	1 1/2 to 2 1/2' long; has alternating black and white rings, the black rings usually being split by red rings.	Found in coniferous forest, hardwoods, oak and brush; usually secretive in habit, so rarely seen.	Found in interior mountains from Douglas Co., Oregon, south.
COMMON KING SNAKE (Lampropeltis getulus).	2 1/2 to 3 1/2'; has creamy or light-colored rings against a brown to black ground color.	Found in most habitats, except water, but prefers forest edges, open brush and oak woods.	From Douglas Co., Ore. south and from Mendocino Co.,Calif. S. on coast.
GOPHER SNAKE (Pituophis catenifer). See illus. p. 103.	2 1/2 to 7' long; a brown and yellowish blotched snake; see full description on page 103.	Found in almost every habitat, but prefers woodland-prairie and brush.	Reported from Olympia area in Wash. From Portland south.
COMMON GARTER SNAKE (Thamnophis sirtalis).	2-4' long. See full description and picture on page 103. Very easily angered.	Common in meadows, marshes, and in forests along streams, and in streams, ponds.	Found in all of region except driest areas.
NORTHWESTERN GARTER SNAKE (Thamnophis ordinoides).	1-2' long; has 3 longitudinal light-colored stripes against dark ground color; belly often red-marked; not easily angered; slow moving.	Common in meadows, forest clearings and along streams in damp localities. Secretive in habit and very docile.	Found in most of region except drier areas.
WESTERN GARTER SNAKE (Thamnophis elegans).	1-3' long. Has a bewildering variety of forms, either striped or spotted; aquatic type has more pointed nose scales. Rarely has red marks.	Terrestrial types prefer meadows, clearings, and brush; aquatic type prefers water and banks of streams.	All of region except extreme W. coast Wash. & central & NW. coast of Ore.

Water
Str.Wd.

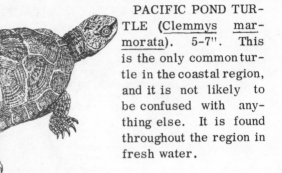

PACIFIC POND TUR-
TLE (Clemmys mar-
morata). 5-7". This
is the only common tur-
tle in the coastal region,
and it is not likely to
be confused with any-
thing else. It is found
throughout the region in
fresh water.

Conif.
Hardw.

NORTHERN ALLIGATOR
LIZARD (Gerrhonotus coerule-
us). 3-5" body and head length.
Alligator lizards are distin--
guished by a fold of skin along the side of the body separating the
dorsal from the ventral scales. The color above is greenish to
bluish with transverse black markings; the ventral color is whitish, with dark lines along the edges of the belly scales.

Conif.
Hardw.

RUBBER SNAKE
(Charina bottae). 1-
2' long. A small rel-
ative of the tropical
boas. The blunt tail,
which looks like the
head, and the uni-
formly brownish up-
perparts are distinc-
tive. It is a secre-
tive snake, usually found underneath boards or logs or hidden in
leaf mold in damp, forested areas. It sometimes actually moves
the tail as if it were a head, so as to confuse an enemy. It hunts
mainly for worms and insect larvae in the leaf mold, but occasionally may catch a mouse. It is one of the most unique and interesting creatures of all the region and should not be harmed in
any way. Like most snakes, save the rattlesnake, it is harmless.

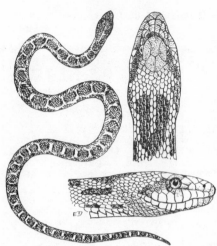

GOPHER SNAKE (Pituo-phis catenifer). 2 1/2 to 8' long. A large, heavy-bodied snake with head not wider than the body. Has light or brownish ground color with large, square to oval dark blotches on the dorsal sur-face; the scales are keeled. This is a harmless snake with a very loud hiss. It does a lot of good by eating harmful rodents and rabbits and should, therefore, be protected. Sometimes rat-tles tail in dry leaves.

Most habitats

COMMON GARTER SNAKE (Thamnophis sirtalis). 2-4' long. This is a snake with yellowish stripes running the length of the body, set off against grayish or brownish background color. Red-dish marks or stripes are often present on the sides. The belly is pale-colored and often bluish. This snake often strikes savagely when attacked or handled and, as with most garter snakes, gives off a bad odor and slimy secretion. Scales keeled. A good swimmer in ponds and streams. Common all over.

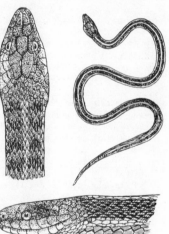

Meadow Str.Wd. Water Strand Conif. Hardw.

CHART OF AMPHIBIAN HABITATS

Name	Description	Habitats	Range
LONG-TOED SALAMANDER (Ambystoma macrodactylum)	2-6' long head and body; head broad and somewhat flattened; a broad greenish-yellow to tan stripe (or blotches) down brown back.	Found under loose rocks or logs or de-bris near lakes or ponds; also found in rotting wood or bark.	From Del Norte Co., California, north to Alas-ka.
NORTHWESTERN SALAMANDER (A. gracile)	2-6" head and body; large glands present on back of head; dark brown body color.	Found on land beneath driftwood, rocks, etc. near streams; in water.	From Sonoma Co., California N. to Alaska.

CHART OF AMPHIBIAN HABITATS (continued)

Name	Description	Habitats	Range
TIGER SALA-MANDER (Ambystoma tigrinum)	2-6" long body and head. Black on back, with numerous spots or bars of yellow; eyes small; head broad and flat.	In dry weather adults live in crevices, rotten logs, animal holes,etc. in open country. In wet weather, found in water or nearby debris.	Santa Cruz, Santa Clara & Monterey Cos., California, in our region.
PACIFIC GIANT SALAMANDER (Dicamptodon ensatus). See illus. p. 106	4-7" body and head. See description and illustration on page 106. The large, thick, spotted body is very distinctive.	In humid coniferous and hardwood forests, in the water or under logs, rocks, bark, etc. near a cool stream.	From Santa Cruz Co., California, north to British Columbia
OLYMPIC SALA-MANDER (Rhyacotriton olympicus).	1 1/2 to 2 1/2" long head and body. A small, slim animal with very short legs and toes; yellow belly; dark mottled above (Ore. and Calif.) or brown with light flecks (Wash.).	Found in or very near swift, cold, shallow water, well-shaded; or under rocks in moss-covered seepages and springs.	Found from Mendocino Co., California, north to Washington.
WESTERN RED-BELLIED NEWT (Taricha rivularis).	2 1/2 to 3 3/4" long head and body. Eyes rather large and protuberant; color usually black to dark brown above, sharply distinct from bright red undersides; eyes dark.	Found in or near to streams in redwood forest; usually found under logs, rocks, boards, debris, etc. Crawls over ground in wet weather.	Found from Sonoma Co., California, north to Humboldt Co. in same state.
ROUGH-SKINNED NEWT (Taricha granulosa). See illustration on page 106.	2 to 3 1/2" head and body length. Light color on head does not reach lower eyelid; brown or black above, orange on the undersides.	Found in humid coniferous and hardwood forests; also in meadows; always near to streams, ponds & lakes; under objects in dry time.	Found from Santa Cruz Co., California, north to British Columbia.
CALIFORNIA NEWT (Taricha torosa).	2 1/2 to 3 1/2" head and body length. Light color on head reaches up to lower eyelid; color reddish-brown or tan above, pale yellowish-orange below.	Found in or near canyon streams, usually associated with oak trees; travels clumsily overland in wet weather. Both adults and young are good swimmers.	From Monterey Co. to Sonoma Co. in California
DEL NORTE SALAMANDER (Plethodon elongatus).	2 3/8 to 3" head and body length. There are 17-19 grooves on the side between the legs. Color usually black or with reddish-brown stripe; belly slate-gray.	Found in or very near old rock slides or piles of rock that are usually damp and covered with moss; generally near to streams or seepages.	Found from Humboldt Co., California, to just north of the Rogue River in Oregon.
DUNN'S SALA-MANDER (Plethodon dunni).	2 1/2 to 3" head and body length. Side grooves 14-16; back stripe yellowish-tan or green.	Found in moss-covered rock piles in shaded seepages near permanent streams.	Found from southwestern Oregon to S.W. Washington

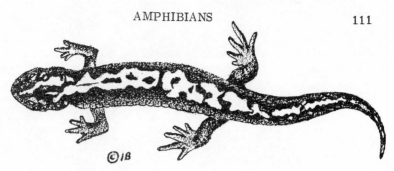

Long-toed Salamander   IRIN BAXTER

California Slender Salamander

Arboreal Salamander

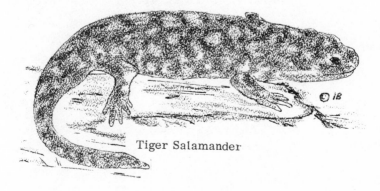

Tiger Salamander

Tailed Frog

Spotted Frog

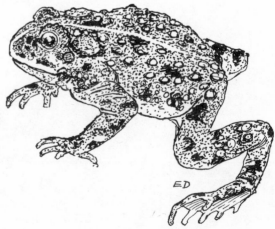

Western Toad

CHART OF AMPHIBIAN HABITATS (continued)

Name	Description	Habitats	Range
WESTERN RED-BACKED SALAMANDER (Plethodon vehiculum).	2 to 2 1/3" head and body length; side grooves 16 or 15; light-colored back stripe reaches to tip of tail, but a few animals all black or all yellowish-orange; belly bluish-gray mottled.	Found in damp coniferous forests or hardwoods, usually under rocks, bark, rotting wood, logs, moss, crevices, etc. and with moist soil.	From Coos Co., Oregon, north to British Columbia.
VAN DYKE'S SALAMANDER (Plethodon vandykei).	2 to 2 1/2" head and body length. A swollen area back of head; side grooves 13 or 14; tan to yellowish back stripe; body brown or black; belly dark with tiny white flecks or spots.	Found under wet or damp rocks near streams; also in mossy rock seepages and under damp bark or surface litter; usually with douglas fir or maples near.	Found in Olympic Peninsula and southwestern Washington.
ESCHSCHOLTZ'S SALAMANDER (Ensatina eschscholtzii).	1 1/2 to 3" head and body length. The only salamander with the tail very narrow at the base. Variable color patterns.	Found in Douglas fir-maple forest; also in redwoods, oaks & brush. In dry periods lives in holes; wet, on surface.	Found in all of region.
CALIFORNIA SLENDER SALAMANDER (Batrachoseps attenuatus).	1 1/4 to 2" head and body length. This and the following are the only salamanders in our region with extremely long, slender bodies and tiny legs.	Found in same habitats as above salamander, but more often under bark and forest debris or leaf mold.	Found from Curry Co., Oregon south through California.
OREGON SLENDER SALAMANDER (Batrachoseps wrighti).	1 1/2 to 1 3/4" long head and body length; very slender & with tiny legs. Conspicuous white spots.	Found associated with red cedar, Douglas fir and maple; usually under rotting wood & bark.	Found only in Oregon Cascades and foothills.
CLOUDED SALAMANDER (Aneides fereus).	2 1/4 to 2 3/4" long head and body. Tips of toes square-tipped and broad for tree climbing; color dark brown, mottled with yellow, gray or whitish.	Found under bark of fir, cedar, redwood, etc. or in leaf litter of tree crotches or hollow stumps; likes recently cleared areas.	Found on Vancouver Island, Willamette Valley, & from Lane Co., Ore. to Mendocino.
ARBOREAL SALAMANDER (Aneides lugubris).	2 1/2 to 3 3/4" head and body length; jaw muscles bulge outward; body is brown, spotted with pale yellow; belly is whitish.	Found under bark, logs, rocks, leaf litter, etc.; also in tree cavities or rotten logs, usually in oak woods.	From southern Humboldt Co., California, south along the coast.
BLACK SALAMANDER (Aneides flavipunctatus).	2 1/2 to 3" long head and body. Toes rounded; belly very dark gray or black; dark above, light speckled.	Found commonly under damp rocks, less commonly under logs or bark. usually on forest edges.	From Santa Cruz and Santa Clara Cos., N. to Humboldt Co.
TAILED FROG (Ascaphus truei)	1-2" long, not including "tail". Outer toe of hind foot broad; pupil vertical.	Lives in or near cold, swift streams under redwood, fir & spruce.	From Mendocino Co. north to Brit. Colum.

Name	Description	Habitats	Range
PACIFIC TREE FROG (Hyla regilla). Illus. on page 107.	2" long or less; the only small frog of region with suction-tip toes. See also page 107.	Found in brush along streams, rock fissures, rodent holes, building crevices; near water.	Found in all of region.
RED-LEGGED FROG (Rana aurora). See illus. on page 107.	About 5" long; almost always with red under legs; a blackish eye-mask reaches from the nose to angle of jaw.	Found mainly in quiet or slow-moving water, but sometimes in near-by ferns and other damp vegetation.	Found all over region near or in water.
YELLOW-LEGGED FROG (Rana boylei).	2 to 3 1/2" long; always colored whitish or cream on belly, changing to yellow on under-side of legs.	Found close to water, preferring creeks with rocky bottoms and with moving (not swift) water.	From upper Willamette Valley & Coos Co. in Oregon south into Calif.
SPOTTED FROG (Rana pretiosa).	3-4" long; very similar to red-legged frog, but smaller and has up-turned eyes.	Found in quiet water; not a strong jumper and can be captured easily.	Found in Willa-mette Valley & S. to Medford area, Oregon.
BULLFROG (Rana cates-biana).	Around 8" or more long; eardrums unusually large. Very large size is distinctive.	Found mainly in quiet water, but sometimes in mountain streams.	Scattered local-ities from San-ta Clara Co., Calif. N. to B.C.
WESTERN TOAD (Bufo boreas).	2 1/2 to 5" long. Body covered with warts.	Found in most habi-tats except dry brush.	Found in all of region.

PACIFIC GIANT SALA-MANDER (Dicamptodon ensatus). 4-7" long. One of the largest of western salamanders. The black blotches on a brown or gray skin are distinctive. Skin not folded on side of body; has strange, barking call. Eats insects, mice.

ROUGH-SKINNED NEWT (Taricha granulosa). 2 to 2 1/2" long head and body. The newt, mud puppy, or water dog, is one of the most commonly observed salamanders. The bright orange-colored under-parts, contrasting with the brown or blackish upper surface, are distinctive. This species usually is found near streams, but, in wet weather, may travel a considerable distance from permanent water.

PACIFIC TREE TOAD (Hyla regilla). 1-2"
long. This small frog is often found well away
from water, and not uncommonly is seen or
heard in bushes or small trees. The black
eye stripe in a frog of this size is distinctive.
The color is variable and tends to resemble
its habitat, which is most generally bright
green. Has toe disks, and a two-syllable,
high-pitched call note, surprisingly loud.

*Water
Str.Wd.
Marsh
Hardw.
Conif.
Brush*

RED-LEGGED FROG (Rana
aurora). Approximately 5" in
length. Usually found near per-
manent streams, ponds, or
lakes, avoiding swift-running
water, where the related yellow-legged frog (page 106) will
often be found. Color brown to
greenish above, with black spots.
Legs with black bars or blotches
and with reddish coloring on the
under-side of the hind legs. It
has a low-pitched, gutteral call
note.

*Water
Marsh
Str.Wd.*

*   *   *   *   *   *

FISHES

If oceans and bays are excluded, there are relatively few species of fish in this region. But the streams of the coast are noted
for their sports fishing. Most anglers seek the salmon and steelhead, which abound in the larger rivers, but others prefer to seek
the cutthroat trout in the smaller coastal streams. Large numbers of salmon and steelhead return each year to the coastal rivers. Here the salmon spawn and die, if they are not caught by
fishermen. The steelhead spawn and return downstream.

Distinguishing features between salmon and trout:

    Salmon: anal fin rays 12 - 19; mouth lining black.
    Trout: anal fin rays 9 - 12; mouth lining white.

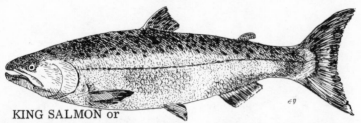

KING SALMON or
CHINOOK (Onchorhynchos tschawytscha). An important sports
fish. The first year of life the young fish move downstream to the
ocean. In the fifth year the adults move up the rivers to spawn.

SILVER SALMON or COHO (Onchorhynchos kisutch).   Distinguished from chinook   by black spotting being confined to back
& upper lobes of tail.   Has needle-like teeth.

STEELHEAD and   RAINBOW TROUT (Salmo gairdnerii).
The steelhead, a famous sporting fish, is simply a sea-running
rainbow trout.   Both lack the red throat mark of the cutthroat.
Dark spots on back large.

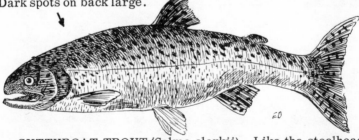

CUTTHROAT TROUT (Salmo clarkii). Like the steelhead the
cutthroat is spawned in fresh water, but spends most of its growing period in salt water. Adults enter streams in autumn or winter. The red streak on each side of lower jaw is a distinguishing
feature; also the long head.   A prized sport fish.

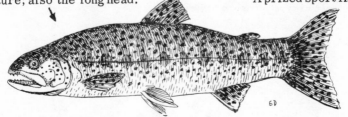

SUGGESTED REFERENCES

Abrams, Leroy, ILLUSTRATED FLORA OF THE PACIFIC STATES. 4 volumes. 1940-1960. Stanford University Press

Bailey, Vernon. 1936. THE MAMMALS AND LIFE ZONES OF OREGON. North American Fauna No. 55. U. S. Department of Agriculture.

Brown, Vinson and Henry G. Weston, Jr. 1961. HANDBOOK OF CALIFORNIA BIRDS. Naturegraph Publishers.

Burt, William H. 1952. A FIELD GUIDE TO THE MAMMALS. Houghton Mifflin Co.

Dalquest, Walter W. 1948. MAMMALS OF WASHINGTON. University of Kansas Publications, Museum of Natural History, Vol. 2.

Gabrielson, Ira N. and S. G. Jewett. 1940. BIRDS OF OREGON. Oregon State College Monographs, Studies in Zoology, No. 2.

Grinnell, Joseph; J. S. Dixon; and J. M. Linsdale. 1937. FUR-BEARING MAMMALS OF CALIFORNIA, their natural history, systematic status, and relations to man. University of California Press (2 volumes).

Hoffman, Ralph. 1955. BIRDS OF THE PACIFIC STATES. Houghton Mifflin, Boston.

Ingles, Lloyd Glenn. 1956. MAMMALS OF CALIFORNIA AND ITS COASTAL WATERS. Stanford University Press.

Jepson, W. L. 1957. A MANUAL OF THE FLOWERING PLANTS OF CALIFORNIA. University of California Press.

Jewett, S. G.; W. P. Taylor; W. T. Shaw; and J. W. Aldrich. 1953 BIRDS OF WASHINGTON STATE. University of Washington Press.

McMinn, Howard E. 1939. AN ILLUSTRATED MANUAL OF CALIFORNIA SHRUBS. J. W. Stacey, San Francisco.

Miller, Alden H. 1951. AN ANALYSIS OF THE DISTRIBUTION OF THE BIRDS OF CALIFORNIA. Univ. of Calif. Publ. Zool., vol. 50, number 6.

Stebbins, Robert C. 1954. AMPHIBIANS AND REPTILES OF WESTERN NORTH AMERICA. McGraw-Hill, New York.

Yocom, Charles F. 1951. WATERFOWL AND THEIR FOOD PLANTS IN WASHINGTON. University of Washington Press.

INDEX

(NOTE: In this index individual species are listed by name only if there is just one of the kind in the book or if the plant or animal belongs to a group, such as the oaks, which covers more than two pages. For other species, look under group names such as "bats", or generic names, such as Quercus or Plethodon, which appear underlined.)

Abies grandis, 28
Abronia latifolia, 12
Accipiter striatus, 81
Acer macrophyllum, 47
Achillea millefolium, 49
Achlys triphylla, 42
Adder's tongue, 39
Agelaius phoeniceus, 98
Alder, red, 29
Alnus rubra, 29
Ambystoma, 110
Amphibians, 109-115
Anaphalis, 43
Anas platyrhynchos, 79
Aneides, 113
Anseriformes, 79
Antrozous pallidus, 74

Aphelocoma californica, 92
Aplodontia family, 68
Aplodontia rufa, 68
Apodiformes, 89
Aquilegia, 41
Arbutus menziesii, 46
Arctostaphylos, 33
Ardea herodias, 78
Arrowhead, 14
Asarum, 40
Ascaphus truei, 113
Aster chilensis, 49
Aster, common, 49
Astur atricapillus, 81
Aythya, 79-80
Azalea, western, 30

Baccharis pilularis, 23
Badger, 64
Bassariscidae, 61
Bassariscus astutus, 61
Batrachoseps, 113
Bats, 74-75
Bear, black, 58
Bear brush, 21
Beaver, 68
Beaver, mountain, 68
Bittern, 78
Blackberry, California, 18
Blackbird, 98
Bluebird, Mexican, 95
Blue-blossom, 20
Bleeding-heart, Pacific, 42
Bobcat, 59

Bombycilla, 97
Bombycillidae, 97
Bonasa umbellus, 83
Botaurus lentiginosus, 78
Brant, black 79
Branta, 79
Brodiaea, 52-53
Brodeaeas, 52-53
Bubo virginianus, 88
Bufo boreas, 114
Bulrush, 15
Bush-tit, 93
Buteo jamaicensis, 81

Cacomistle family, 61
Calamagrostis nutkaensis,53
Calypte anna, 89
Canis latrans, 59
Capella gallinago, 85
Caprimulgiformes, 88
Carnivores, 58
Carpodacus, 99
Cascara, 30
Casmerodius, 78
Castanopsis, 48
Castilleia latifolia, 22
Castor canadensis, 68
Castoridae, 68
Cat family, 58
Cat, ring-tailed, 61
Cat-tail, 15
Cathartes aura, 81
Catoptrophorus semipal-
    matus, 85
Cedar, giant, 29
Ceanothus, 20, 31
Certhia familiaris, 94
Cervus canadensis, 56
Chaetura vauxi, 89
Chamaea fasciata, 94
Charadriiformes, 84
Charadrius vociferous, 84
Charina bottae, 104, 108
Chickadee, chestnut-backed,
    93
Chickaree, 66
Chinook, 116
Chinquapin, 48
Chipmunk, 67
Chiroptera, 74
Chordeiles minor, 88
Chrysemys picta, 104
Ciconiiformes, 78
Cinclus mexicanus, 95
Circus cyaneus, 82
Citellus, 67
Clemmys marmorata, 104,
    108
Clethrionomys californicus,
    71

Clintonia, 39
Clintonias, 39
Coho, 116
Colaptes cafer, 90
Coltsfoot, 39
Coluber constrictor, 107
Columba fasciata, 87
Columbiformes, 87
Colybiformes, 77
Columbines, 41
Colymbus auritus, 77
Contia tenuis, 107
Contopus richardsonii, 91
Coot, 84
Coraciiformes, 89
Corvus, 93
Corynorhinus rafinesque, 74
Coyote, 59
Coyote brush, 23
Cranes, 84
Creeper, 94
Creeper family, 94
Crossbill, red, 100
Crotalus viridis, 104-105
Crow, 76, 93
Crow family, 92
Cudweed, lowland, 44
Cupressus macrocarpa, 9
Curlew, long-billed, 85
Currants, 33
Cyanocitta stelleri, 93
Cypress, 9

Dafila acuta, 7
Daisy, seaside, 22
Danthonia californica, 53
Deer, 56-57
Deer family, 56-57
Deer brush, 31
Deer-foot, 42
Dendragapus obscurus, 83
Dendrocopus, 90
Dendroica, 98
Diadophis amabilis, 106, 107
Dicamptodon ensatus, 110,
    114
Dicentra, 42
Dicentras, 42
Dipper, 95
Disporum smithii, 36
Distichlis spicata, 14
Dog family, 59
Dove, mourning, 76, 87
Dowitcher, 86
Dryocopus pileatus, 90
Ducks, 79-80

Eelgrass, 14

Egrets, 78
Elk, Roosevelt, 56
Empidonax difficilis, 91
Ensatina eschscholtzii, 113
Epilobium, 48
Eptesicus fuscus, 74
Erigeron glaucus, 22
Eriophyllum staechadifolium,22
Eumeces skiltonianus, 104
Euphagus cyanocephalus, 98
Eutamias, 67

Fairy lantern (or bell), 36
Falconiformes, 81
Falcon, peregrine, 82
Falco, 82
Felis concolor, 58
Fern, sword, 25
Finch family, 99
Finches, 99
Firs, 24, 27, 28
Fire-cracker plant, 53
Fire-weed, 48
Fisher, 62
Flicker, red-shafted, 90
Flycatcher family, 91
Flycatchers, 91
Foxes, 60
Fragaria chilensis, 11
Frogs, 113, 114, 115
Fulica americana, 84

Galliformes, 83
Garrya, 21
Gaultheria shallon, 34
Gavia, 77
Gaviiformes, 77
Geese, 79
Geothlypas trichas, 97
Gerrhonotus, 104, 108
Ginger, wild, 40
Glaucomys sabrinus, 66
Gnaphalium, 44
Goldenrod, coast, 12
Goldfinch, American, 100
Gooseberry canyon, 32
Gophers, 67
Goshawk, 81
Grape, Oregon, 35
Grasses, 53
Grebes, 77
Grosbeaks, 99
Grouse, 83
Gruiformes, 84
Gulls, 86

Habitats, types, 5
Hare, snowshoe, 65
Hare family, 65
Hawks, 81-82

Hemlock, western, 26, 28
Heracleum lanatum, 17, 18
Herons, 78
Hesperiphona vespertina, 99
Himalaya berry, 19
Hirundinidae, 92
Holcus lanatus, 53
Honeysuckle, wedded, 21
Huckleberries, 34
Hummingbirds, 89
Hyla regilla, 114, 115
Hylocichla, 96

Icteridae, 98
Indian lettuce, 35
Insectivores, 73
Inside-out flower, 41
Ixoreus naevius, 96

Jackrabbit, black-tailed, 65
Jays, 92, 93
Junco, Oregon, 101
Junco oreganus, 101

Killdeer, 84
Kingfishers, 89
Kinglets, 96

Lagomorpha (order), 65
Lampropeltis, 106, 107
Larus, 86
Lasionycteris noctivagans, 74
Lasiurus cinereus, 74
Laurel, California, 47
Leporidae, 65
Lepus, 65
Lilium, 37
Lilies, 37
Limnodromus griseus, 86
Linnet, 99
Lion, Mountain, 58
Lithocarpus densiflorus, 47
Lizards, 104, 105, 108
Lizard tail, 22
Lonicera, 21
Loons, 77
Lophortyx californica, 84
Loxia curvirostra, 100
Lupines, 10
Lupinus, 10
Lutra canadensis, 63
Lynx rufus, 59
Lysichiton americanum, 15

Madrone, 46
Mahonia nervosa, 35
Maianthemum dilatatum, 39
Mallard, 76, 79

Manzanita, hairy, 33
Maple, big-leaf, 47
Mareca americana, 80
Marten, 62
Martes, 62
Martin, purple, 92
Megaceryle alcyon, 89
Melospiza, 102
Mephitis mephitis, 64
Merganser, American, 76
Mergus merganser, 76
Mice family, native, 65
Microtus californicus, 67
Mimulus, 22, 53
Mink, 62
Mole family, 73
Moles, 73
Monkey flower, bush, 22
  common, 53
Montia sibirica, 35
Mountain beaver, 68
Mouse, harvest, 70, 71
  house, 55
  jumping, 70, 71
  meadow, 70, 71
  red-backed, 70, 71
  red tree, 70, 71
  white-footed, 70, 71
Muskrat, 69
Mustelidae, 63, 64
Mustela, 63
Myotis lucifugus, 74
Myrica californica, 18
Myrtle, Oregon, 47
  wax, 18

Neotoma, 69
Neurotrichus gibbsi, 73
Newts, 110, 114
Nighthawk, Pacific, 88
Nuphar polysepalum, 16
Numenius americanus, 85
Nuthatch family, 94
Nuthatches, 94
Nuttallornis borealis, 91
Nycticorax nycticorax, 78

Oak, black, 51
  canyon, 29
  Garry, 51
  Oregon white, 51
  tan, 46
Odocoileus, 57
Onchorhynchus, 116
Ondatra zibethica, 69
Oreortyx picta, 83
Otter, river, 64
Otus asio, 87
Owls, 87, 88
Oxalis oregana, 36
Oxyura jamaicensis, 80

Painted cup, seaside, 22
Paridae, 93
Parsnip, cow, 17, 18
Parulidae, 97
Parus rufescens, 93
Passerculus sandwichensis, 102
Passerella iliaca, 102
Passeriformes, 91
Pearly everlasting, 43
Perisoreus canadensis, 92
Peromyscus maniculatus, 71
Pewee, western wood, 91
Phenacomys longicaudus, 81
Pheucticus melanocephalus, 99
Phrynosoma coronatum, 104, 105
Picea sitchensis, 28
Piciformes, 90
Pickle-weed, 14
Pigeon, band-tailed, 87
Pines, 8
Pine siskin, 100
Pinus, 8
Pipilo, 100
Piranga ludoviciana, 99
Pituophis catenifer, 107, 109
Plant parts, 6
Plethodon, 110, 113
Podilymbus podiceps, 77
Procyonidae, 60
Procyon lotor, 61
Progne subis, 92
Psaltriparus minimus, 93
Pseudotsuga menziesii, 27

Quail, 83, 84
Quercus, 29, 51

Rabbit, brush, 65
Raccoon, 55, 61
Raccoon family, 60
Rana, 114, 115
Rat, wood, 69
Raven, 93
Redwood, 27
Regulus, 96
Reithrodontomys megalotis, 71
Rhamnus purshiana, 30
Rhododendron, California, 30
Rhododendron, 30
Rhyacotriton olympicus, 110
Ribes, 32, 33
Robin, 95
Rodents, 66
Rose bay, 30
Rubus parviflorus, 32
  spectabilis, 19
  thrysanthus, 19
  vitifolius, 18

Saggitaria, 14

Salal, 34
Salamander, arboreal, 113
  black, 113
  clouded, 113
  Del Norte, 110
  Dunn's, 110
  Eschscholtz's, 113
  long-toed, 109
  northwestern, 109
  Olympic, 110
  Pacific giant, 110, 116
  slender, 113
  tiger, 110
  Van Dyke's, 113
  Western red-backed, 113
Salicornia, 14
Salmo, 116
Salmon-berry, 19
Salmon, 116
Salt grass, 14
Sapsucker, yellow-bellied, 90
Scapanus, 73
Sceloporus occidentalis, 104
Sciuridae, 66
Sciurus griseus, 66
Scoliopus, 39
Selasphorus, 89
Sequoia sempervirens, 27
Shrew family, 73
Shrews, 72, 73
Shrew-mole, 72, 73
Sialia mexicana, 95
Silk tassel, 21
Sitta, 94
Sittidae, 94
Skink, western, 104, 105
Skunks, 64
Skunk cabbage, 15
Slinkpod, 38, 39
Smilacina racemosa, 40
Snake, garter, 107, 109
  gopher, 107, 109
  king, 106, 107
  racer, 106, 107
  rattle, 104, 105
  ring-necked, 106, 107
  rubber, 104, 108
  sharp-tailed, 107
Snow bush, 31
Snipe, Wilson's, 85
Solidago spathulata, 12

Solomon's seal, 39, 40
Sorex, 73
Soricidae, 73
Sorrel, redwood, 36
Sparrows, 101, 102
Spatula clypeata, 79
Sphyrapicus varius, 90
Spilogale putorius, 64
Spinus, 100
Spizella passerina, 101
Spruce, Sitka, 24, 28
Squaw mat, 31
Squirrel family, 66-67
Squirrels, 66-67
Star flower, 44
Steelhead, 116
Strawberry, sand, 11
Strigiformes, 87
Strix occidentalis, 88
Swallow family, 92
Swallows, 92
Swans, 79
Swift, Vaux, 89
Sylvidae, 96
Sylvilagus bachmani, 65
Tachycineta thalassina, 92
Talpidae, 73
Tamiasciurus douglassi, 66
Tanager, western 99
Tanager family, 99
Taricha, 110, 114
Taxidea taxus, 64
Thamnophis, 105, 107
Thimbleberry, 34
Thomomys, 67
Thraupidae, 99
Thrush family, 95
Thrushes, 95, 96
Thryomanes bewicki, 94
Thuja plicata, 29
Titmouse family, 93
Toad, western, 114
Tobacco bush, 31
Towhees, 100
Trientalis europaea, 43
Trillium, 36, 37
Trilliums, 36, 37
Troglodytidae, 94
Troglodytes troglodytes, 95
Trout, 116
Tsuga heterophylla, 30

Turdidae, 95
Turdus migratorius, 95
Turtles, 104, 106
Typha latifolia, 15
Twinberry, black 21
Umbellularia californica, 47
Urocyon cinereoargenteus, 60
Ursidae, 58
Ursus americanus, 58
Vaccinium, 34
Vancouveria planipetala, 41
Vancouveria hexandra, 42
Vancouveria, small-flowered, 42
Vermivora celata, 97
Vespertilionidae, 74
Verbena, sand, 11
Vireonidae, 97
Vireo family, 97
Vireo, Hutton's, 97
Vireo huttoni, 97
Voles (meadow mice), 70, 71
Vulpes fulva, 60
Vulture, turkey, 81
Wake-robin, Western, 39
Warbler family, 97
Warblers, 97, 98
Water lily, yellow, 16
Waxwing family, 97
Waxwings, 97
Weasel family, 62-64
Weasels, 63
Willet, 85
Willow herb, 49
Wilsonia pusilla, 97
Woodpeckers, 90
Wren family, 94
Wrens, 94, 95
Wren-tit family, 94
Wren-tit, 94
Yarrow, 49
Yellowthroat, 97
Zapodidae, 71
Zapus princeps, 71
Zenaidura macroura, 87
Zonotrichia, 101
Zostera marina, 14